t· ·)ue

Histological Techniques *for* Electron Microscopy

Histological Techniques *for* Electron Microscopy

DANIEL C. PEASE

Department of Anatomy
University of California School of Medicine
Los Angeles, California

1960

ACADEMIC PRESS *New York and London*

ACADEMIC PRESS INC.
111 FIFTH AVENUE
NEW YORK 3, N. Y.

United Kingdom Edition
Published by
ACADEMIC PRESS INC. (LONDON) LTD.
17 OLD QUEEN STREET, LONDON S.W. 1

Library of Congress Catalog Card Number 60-8052

PRINTED IN THE UNITED STATES OF AMERICA

PREFACE

This is intended to be a very practical book for those who would study cells or tissues with an electron microscope. Although all the tools have been at hand for this sort of work for more than five years, effective work has been done in relatively few laboratories. In general this is because of the complexity of the techniques and the consequent lack of fully trained investigators; yet there is no one aspect of the problem that is particularly difficult in itself. But success depends upon attention to much detail, as well as knowledge and competence in widely diversified fields. The adage that "a chain is no stronger than its weakest link" is quite pertinent. To make successful electron micrographs of tissues one must preserve the specimens with skill, section them with art, use an electron microscope with understanding, and do photography with facility.

Except for a few "old timers" who grew up with the subject, most senior investigators who are doing effective work today apprenticed themselves at one of two laboratories, in the United States, at the Rockefeller Institute, in Europe, at the Karolinska Institutet. Of course with time new clones have been established, and there are now a number of "teaching laboratories" to which one can go for instruction and experience. No doubt, a year's apprenticeship in such a laboratory can be the most efficient, effective, and stimulating way of becoming established in this field. But there are many who are interested, yet for various reasons find it impractical or impossible to serve this apprenticeship. It is to this group in particular that this volume is directed.

It should be possible for a novice to gain reasonable competence through his own ingenuity if he has a guidebook to help with the maze of detail. It will assist the beginner immensely if he can arrange to spend two or three days with someone competent in this field who will systematically go through all of the steps involved with him while he takes notes. This will, in essence, be a survey which will get the novice started. He should then go back to his own laboratory and work for a few weeks, or even months, as best he can with books and with whatever local help is available. When he has done enough work to understand his own limitations and special difficulties, then it is time for him to return to the experienced laboratory for an "advanced course." Then much can be accomplished in even a short one-day visit. There will come the time when the beginner qualifies as an investigator, but he should continue to seek out his peers and refine his technique. This, of course, is a never ending effort for all of us, but when the investigator is sufficiently advanced to know what questions to ask it is not difficult to improve rapidly.

Difficulties in getting started in this field result, in large measure, from the novice's overlooking seemingly small, but actually critical details. This book therefore makes an effort to present reasons for doing things in particular ways in the hope that the details then will seem significant. Sometimes, no doubt, this results in wordy overemphasis. In spite of good intentions on the part of this author, undoubtedly some facets of the work will not be understood or absorbed easily. When in trouble, the novice is urged to return to the pertinent parts of the text to search for overlooked details.

Difficulties in getting started also result from the necessity of performing all sorts of essential activities on a much finer scale than most people have had practice with. A jeweler's work is coarse by comparison. The unaided eye is quite inadequate. This author uses binocular head loops of varying magnifications for all sorts of tasks including tissue preparation, some aspects of section mounting, observation of the electron microscope's fluorescent screen, and examination of negatives. He is as adjusted to these magnifying aids as to an ordinary pair of glasses. In addition, a variety of activities require the magnifications of binocular "dissecting" microscopes. These include sectioning, section mounting, block trimming, focusing of the electron microscope, cleaning polepieces, and preparation and insertion of apertures. Working with one's hands under such a microscope must become facile. This writer is greatly indebted to his first mentor in graduate school who set him to work cutting sea urchin eggs into halves and quarters, freehand. In retrospect this was an excellent exercise in manual dexterity and the use of magnifying aids. From the beginning the novice must realize that he must acquire this facility, and he should deliberately develop it at every opportunity.

Some of the barriers to this world of the small and delicate are psychological rather than physical. The person who has been dissecting with stiff blunt forceps, and who has been trimming paraffin blocks with a dull scalpel, is going to have to refine his tools as well as his movements, and it may prove harder for him than for an individual without experience with conventional microtomy.

This book deliberately avoids being a scholarly presentation. Credits ordinarily are coupled only with major developments in the field. Partly this is because it would be difficult indeed to unravel the origin of every bit of technique. One often hears of new "tricks" by word of mouth, perhaps many times removed from the source. This writer knows full well that he depends importantly on the ingenuity of many in ways unknown to him.

This volume avoids the scholarly approach also in making no effort to detail all technical proposals of the literature. There has been so much interest in electron microscopy that its bibliography already is enormous. However, as far as technique is concerned in this field which has expanded so rapidly, what has seemed important one year often has dwindled to insignificance the next. Although this writer is aware that a few individuals will object violently to aspects of technique as they are presented in this book, an effort has been made to be conservative and to present what is really useful and enduring. In addition, certain new and exciting developments are presented as alternates, but it is hoped with the clear understanding that they are still experimental and may have only limited usefulness. Above all else, a method is presented which the author knows to be workable, and which is certainly adequate to get a beginner started. There will be time enough for the novice to develop refinements or alternates later.

This writer has profited greatly from knowing most of the American and a few of the European electron microscopists who were involved from the first in developmental work with ultrathin sections. These investigators will recognize the fruits of their own ingenuity here and there in this text, and perhaps this book will indirectly pay the author's debt to these individuals by fostering an interest in what is, after all, our common scientific enthusiasm. This writer also is happy to acknowledge that he owes much to a small group of visiting investigators with varying experience who have worked in his laboratory, and to a group of graduate students past and present, all of whom have been inventive.

This author's laboratory could not have been established and maintained without financial support, first from the Los Angeles Heart Association, and later from the United States Public Health Service. Thus their share in this book is very real.

Los Angeles, California DANIEL C. PEASE
March 1960

CONTENTS

1. Organization and Management of an Electron Microscope Laboratory

1.1. Laboratory Layout

The requirements for an electron microscope laboratory can be quite accurately defined if it is to be used almost exclusively for tissue work. There must be a microscope room which can be darkened, a photographic darkroom as well, and a preparation room. Although an office area may be combined with the latter, it is highly desirable to have an additional separate room for paper work.

Electron microscope manufacturers will supply a customer with detailed specifications of the requirements for their product. The minimum shape and size of the microscope room depends partly upon the shape of the microscope, and also upon the clearances that are required to open the instrument for service. The latter are apt to be overlooked or forgotten by one without experience. For this reason a microscope ordinarily cannot be backed up against a wall or be placed in too narrow a closet. It will probably have to stand more or less in the center of a modest sized room.

Most manufacturers stress the need for a stable base for their product, although instruments differ in their sensitivity to vibration. This is the reason why so many electron microscope laboratories are to be found in basements. Elevator shafts in the vicinity are to be avoided particularly. Large transformers, generators, motors, and the like have powerful magnetic fields about them that exert their influence for many feet. They may disturb the operation of an electron microscope in an adjoining room.

The manufacturer will specify what special utilities must be brought into the room. These will probably include water to cool a diffusion pump, as well as one or two extra electrical circuits. Some microscopes use massive voltage stabilizers in their power line. These have powerful magnetic fields about them, and therefore cannot be located within perhaps 20 ft of the microscope, but they can be hung outside of the room in a hall, or perhaps be put in the preparation room.

The microscope room must be in essence a darkroom, although not as light tight as a photographic darkroom. It is essential that the room have some sort of forced ventilation for a microscope liberates a good deal of heat, and the operator will spend hours of time in the room with the door closed. The photographic darkroom should adjoin the microscope room so that the operator can go in easily to unload and load photographic plates. He may also wish to develop plates as they are exposed to keep track of what he is doing, so the darkroom will be in constant demand whenever the microscope is being operated.

The darkroom need not be elaborate, for the photographic requirements of electron microscopy are specific and limited. Darkroom mess mostly can be eliminated if photographic development is carried on within a very large sink. This should be big enough to accommodate at least four 11 × 14 in. trays. Even this assumes that a circulating tank for print washing stands outside of the sink and simply drains into it. At least two faucets should be provided, one of which is used for print washing. In addition there must be bench space for a good photographic enlarger and its easel, as well as bench space to lay out enlarging paper and for a print dryer.

An active electron microscope laboratory certainly should have its own darkroom and not be required to share these facilities with other groups. The darkroom will be in such constant demand that others using it will not only create annoyance, but seriously impair efficiency.

The preparation room will serve a certain number of quite specific functions, and it can best be arranged so that each of these is carried on in its own particular place. The principal activities will include; (1) sectioning, with a microtome and its associated equipment, (2) tissue preparation, with space for small animal dissection, (3) chemical work, with a balance and a pH-meter where solutions can be prepared, (4) general purpose light microscopy, (5) tissue dehydration requiring a sink, (6) "dirty work" including glass washing, with a sink and preferably with a table topped with asbestos board and equipped with a Bunsen burner where soldering operations, minor mechanical repair work and simple glass blowing can be carried on, (7) several kinds of storage areas, including shelving for chemicals and pieces of equipment, and drawers for small objects including dissection tools. In addition, there should be bench space for ovens and an ultraviolet lamp which will be used for plastic polymerization. There must be floor space for an evaporating (shadow-casting) unit, and for a refrigerator. It will be difficult indeed to carry on all of these activities with less than 50 linear feet of bench space. This means a room of about 200 sq ft. If office activities have to be incorporated, it must be decidedly larger.

The sizes of the rooms, other than office areas, are not particularly dependent upon the number of people who work in them; although if more disciplines are involved than just tissue work, preparatory techniques become more complex and varied, and consequently require more room. It is better, then, to break up the activities into at least two rooms. The requirements for office space, however, are quite dependent upon the number of people who work in the laboratory. Not only will each person need desk space, but there should be at least one large table where prints can be sorted, trimmed and mounted. This should be equipped with a large cutting board, and perhaps a dry mounting press. There should be floor space for filing cabinets which will store prints and negatives.

It will be apparent from the above that a "small" microscope laboratory conservatively will have to occupy about 500 sq ft. This figure must be substantially increased if there will be more than two or three active investigators, or if it is expected to grow so that there will be more than one microscope, or if investigators representing several disciplines, each with its own technical problems, are to be represented.

1.2. Equipment

The equipment required to do tissue work with an electron microscope can be fairly exactly specified and its cost determined. Since there is some latitude, items will be specified either as essential or desirable. In preparing these lists items valued at less than $50 have not ordinarily been included.

For section work of current standards the electron microscope used must incorporate two features. There must be an externally adjustable objective aperture, and there must be electrostatic compensation of the objective polepiece (or the magnetic equivalent which can be controlled by external means while the microscope is in operation). Although most of the larger microscopes being made today incorporate these features, certain simplified instruments of limited usefulness do not. The older RCA EMU-1 and -2 series instruments, and the newer EML, can be modernized with accessory equipment of this sort purchased from the Canal Instrument Company of Bethesda, Maryland. The cost is about $1000 and $1200, respectively, but is an investment that must be regarded as essential for tissue work. In addition, any microscope used must have a magnifying viewer for accurate focus. This should be a ×7 or ×10 power microscope with a very long working distance. A binocular dissecting microscope head is most satisfactory if it can be fitted to the instrument.

Almost all tissue studies with ultrathin sections can be done very satisfactorily with an electron microscope operating at an accelerating potential of 50 kv. Indeed the reduced contrast resulting from the use of higher voltages generally works to one's disadvantage. There are problems with some of the newer embedding media, however, (see, Sections 4.7, 4.8, and 5.18), which can be partly solved by using fairly thick sections with greater accelerating voltages. Thus, one cannot categorically say that high voltage microscopes will not prove importantly useful in the future. It should be noted, however, that one cannot easily switch back and forth from one accelerating voltage to another without correcting alignment and compensation, so that in practice a microscope almost has to be used for extended periods of time at a particular voltage.

It is desirable that the electron microscope have certain features that may only be available as accessories. An externally adjustable condenser aperture is useful since apertures of various sizes can be substituted easily to meet varying requirements. The electron microscope must have a fast shutter mechanism of some sort (so that 0.5–1 sec exposures can be made). It is desirable that this be an electrically timed shutter. A photometer can be added as an accessory to some microscopes and is a useful adjunct. A microscope equipped with a modulating and normalizing system for eliminating lens hysteresis is useful, particularly if accurate mensuration is anticipated.

Ultramicrotomes are discussed in Chapter 5. It is essential that any instrument used be fitted with a binocular dissecting microscope. This can be mounted on a separate stand with a long horizontal arm, or it can be incorporated into the microtome on a suitable bracket.

At least one additional binocular dissecting microscope must be put on the essential list. It is important that its head be tiltable in relation to its stage so that sections can be examined obliquely. This is so used to observe interference colors which are seen with suitably reflected light, and which are a measure of section thickness.

An evaporating unit is on the essential list. Commercially available units with a full complement of vacuum gauges, valves, and pumps cost about $2000. If the use of the instrument will be only for carbon evaporation for specimen supporting films, such an elaborate system is not truly necessary although it is convenient. A high quality mechanical pump with a bell jar, homemade baseplate, and carbon holder can suffice, and can be put together for a few hundred dollars even if a gauge is included.

A good photographic enlarger with a high quality, fast lens is an important investment. One should not try to save money on this item. A number of modern electron microscopes offer a choice of plate size. The enlarger lens and condenser system that works well with miniature negatives will not also serve with large sizes or vice versa. Thus, if the enlarger is to accommodate both small and large negatives, two different lenses must be purchased as well as two different condenser systems. Enlargers which rely on diffusing systems rather than condenser lenses should be avoided for they do not produce maximum contrast.

An electric print dryer is necessary to handle the volume of prints that may be expected. A good analytical balance, pH meter, laboratory oven, and refrigerator must also be put on the essential list.

As far as desirable equipment goes, at least one further binocular dissecting microscope is very useful, particularly if three or more people are working in the laboratory more or less simultaneously. There are endless minor tasks which are best performed with its aid: trimming specimens, trimming blocks, examining apertures and polepieces for contamination, surveying specimen screens, picking up sections on screens, examining negative detail, as well, perhaps, as delicate operations on small animals. If there is only one general purpose instrument available, this can present an annoying and inefficient bottleneck.

An ordinary compound microscope is occasionally important, particularly for studying and measuring apertures, although much of the time it may stand idle. Certain types of work, however, require a close correlation between ordinary light microscopy and electron microscopy, and then such an instrument may be used regularly. A phase contrast microscope is also desirable, for with it one can check on the contents of ultrathin sections as they are being cut by taking a sample section 0.5–1 μ thick, and looking at it directly after covering it with immersion oil.

In the darkroom a photoelectric densitometer for judging the exposure for enlarging is useful.

A small electronic "shop" is almost essential. Even the complete novice will, on occasion, have to wield a soldering iron, and he should soon learn to use a volt-ohmmeter. If he is or becomes more knowledgeable, he will want additional test equipment including, probably, a tube checker and a cathode ray oscilloscope.

A small machine shop is also almost inevitable. No doubt this will begin with a collection of hand tools and a vise. There are many times when an electric hand drill is most useful. If a small lathe with a milling attachment, a drill press, and an electric grinder are available, these will occasionally prove of great value if the investigator has any inventive propensity.

1.3. Management of the Laboratory

Upper echelon administrators like to think they are being efficient if they can get several diverse groups to share common facilities. They are apt not to realize that for every dollar of equipment money that they may save this way they may lose many times this in salary money. After all, if the laboratory is unproductive, the capital investment has been wasted. The administrator with no particular knowledge of the problems involved often assumes that a single electron microscope laboratory can be run as a service facility for a whole school or university, or for a complex industrial plant with diverse interests. At first thought this does not seem unreasonable, yet this implies that the responsible person in the laboratory be a technician for every physicist, chemist, engineer, bacteriologist, virologist, zoologist, botanist, pathologist, or anatomist who may walk in with a naive notion of a problem. No truly able, independently minded, qualified electron microscopist will put up with this for long. His energy and originality would be dissipated in unfruitful directions. It is an intolerable situation for a person of advanced training and highly technical knowledge. In general, "service laboratories" do not work and cannot be expected to work!

Typical of this common administrative attitude was a grant application to a federal agency for funds to establish an electron microscope laboratory which was reviewed by this writer. The application came from a university, and outlined over 20 separate projects by as many different investigators, in at least a dozen different departments, all of which supposedly were to go on simultaneously. It was, of course, obvious that all of the scientific departments in the school had been circularized and urged to submit something reasonable that might bolster the application. So far as this author knows, only one of the names on the entire list had the slightest knowledge of electron microscopy. He shudders to think what would have happened if this idealized program actually had been initiated. Fortunately the granting agencies of the United States Government, which have been involved in medical and biological support programs, seem to have been well advised about electron microscopy. They have been quite liberal in giving away microscopes, but only when they could be sure that they would be operated by properly trained and qualified individuals. Their attitude has been "find the man, and we'll give you the scope."

It is only under the most extraordinary of circumstances that a "service laboratory" can turn out a volume of quality work. This is because the man who operates the microscope must know what he is doing at all levels of the investigation. Most particularly, he must be able to form technically sophisticated opinions about what he sees on the fluorescent screen. In histological or cytological study, trained professional judgment is essential in the selection of suitable fields to be photographed. The investigator himself must spend hours that stretch out to days and months at the microscope. This cannot be a part-time job if the problem has any complexity at all. Successful electron microscopy is a very individualistic discipline which inevitably represents much effort on the part of the investigator rather than being delegated work. This is particularly true in tissue study where every field of view differs from the ones before in a constantly variable panorama. The electron microscope laboratories which are deservedly famous for their work with tissues are all built around individuals, or groups of individuals, who are competent investigators in their field of choice, as well as being excellent technicians at the instrumental level. They explore their own interests and develop their own programs.

There are several distinct groups of biological scientists who are interested in tissue morphology. Their disciplines are sufficiently close so that they might efficiently share a common laboratory facility, but as soon as one starts even to add bacteriology, complications begin. The preparatory techniques of the bacteriologist are quite different from the histologist, although these, of course, can be carried out in separate quarters. More serious is the problem of the microscope if it is being shared. The bacteriologist is apt to be working with metal shadowed material, and therefore he does not want an aperture in the objective polepiece. The person working with tissues requires the objective aperture. In most microscopes it is not too easy to switch and then recompensate as certainly would prove necessary. The histologist spends much of his time doing "survey" work where great resolution is not required. Thus the microscope operates for so much time that contamination builds up, and in essence the microscope does not stay "in tune." This is unsatisfactory to the bacteriologist whose specimens are pretty much alike all over, and who does not have to search for hours to find out what he has, or to take a representative series of micrographs. He does want, though, a few pictures with very high resolution. Thus, the problems of the bacteriologist and histologist are sufficiently different so that each is annoyed by what the other does, and there is real inefficiency introduced if they must work with common facilities.

1.4. Service Personnel

Companies which manufacture electron microscopes have made efforts to establish service departments which make the initial installation, train investigators in the operation of the microscope, and can more or less maintain the instrument in operating condition. One cannot assume, however, that because he has a service contract all difficulties will be automatically taken care of. Service personnel are limited by human frailties as well as in their training and their interest in the job. In some areas where there are many instruments of a single manufacturer, microscope maintenance is a full time occupation for one or more well trained electronic engineers. In other areas microscopes are only a part of the business of the service man (in Los Angeles, service personnel have at various times also had to occupy themselves with commercial television installation, military work, weighing sausages, and the problems of bottling beer). In other parts of the world there are, in effect, no available service engineers after the microscope is once installed. Even under the best of circumstances, the director of the laboratory must assume much responsibility himself. There are a number of maintenance tasks such as filament changing, compensation, aperture centering and stability checking which he must perform himself, or delegate to knowledgeable individuals who are actively using the microscope. In general, the director of the laboratory must know when service work is required and seek help as needed. It is not enough to assume that a service contract will guarantee an instrument that performs well. Thus there is no substitute for an able director.

Fortunately many, if not most of the present day electron microscopes, are quite reliable instruments once they have been put in proper working order. After the initial "break-in" period, the troubles should be sporadic, and result from the failure of a particular part, perhaps a tube, condenser, resistor, transformer, or a battery, or, on the mechanical side, a gasket. Laboratories which must be self-sufficient, often can be for very long periods of time. They may get along by recruiting local technical help. Someone on the physics faculty, an electrical engineer down the hall, or even an amateur radio ham often can work out electronic difficulties if the laboratory director can make clear what the problem is. Thus, the director should know enough about the theory and practice of electron microscopy to, (1) be aware of trouble, (2) be able to perform something of a differential diagnosis and, (3) be able to explain it in intelligible terms to someone not specifically trained in the details of the instrument. To a large extent these abilities come with experience, but also the manufacturer provides personal instruction which should be taken advantage of to the fullest possible extent, and there are technical instrument manuals to assist the novice. On the one hand the task of maintaining a microscope should not be forbidding, but there is no substitute for experience and knowledge.

1.5. Technical Help

It is not usually possible to rely heavily on technical help in an electron microscope laboratory. There are certain routine tasks that can be delegated but, in general, these are not the serious bottlenecks. The time consuming operation is the analysis of the material by the responsible investigator, and there is no substitute for hours of personal time at the microscope. The responsible investigator will select the fields to be photographed and do that himself.

He may train a technician to cut and mount sections. However, this is apt to prove frustrating for a variety of reasons. Much time can be wasted at the microscope if the specimen grids are not well covered with excellent material. In this author's experience it is often maddening to try to work with somebody else's material. One has the feeling that he can do better himself, and gets impossibly irritated by too many blank squares, folds in the sections, dirt sprinkled over the surface, "chatter," unsuitable thickness, poor preservation, etc. If the responsible investigator has himself prepared the specimen from the beginning, he has no one to blame but himself for the various flaws which appear so commonly under the best of circumstances. He presumably profits by his experience, but it is difficult indeed to put up with someone else's errors and carelessness. Then, too, there is little chance of obtaining a qualified technician without training him or her in the first place, which is sure to be a time consuming process and one without assurance of success or promise of permanence. It thus seems best that any investigator in this field preserve his own material, section and mount it, and do his own microscopy. The technician is left with certain secondary tasks in accordance with his or her skills. These can include photographic printing and some microscope maintenance, as well as the more prosaic animal care, solution making, bottle washing, and laboratory clean-up.

1.6. Shared Facilities

How many people can or should share the use of a microscope? If the laboratory is entirely concerned with tissue work, there cannot be very many people working actively at the same time without extraordinary inconvenience. It is not unreasonable for an investigator to spend about one third of his actual working time at the microscope. The other two thirds will be spent in specimen preparation, photographic enlarging, library work, etc. Thus, three full-time investigators per microscope probably is most efficient, particularly as some potential microscope time inevitably is lost in its maintenance and repair. A university departmental laboratory with a faculty director and two or three advanced graduate students represents a very good working combination. Since not all of the people will be equally skilled, the director will have enough trouble guiding the activities of two or three people and supervising the maintenance of the microscope for them.

In some laboratories an effort is made to schedule the microscope time. Although this may be a desperate resort when the microscope is being overwhelmed, the method is basically inefficient. Who can say how long it will take an investigator to look at sections from even a single block, and assess and record what he sees? If one is testing a preparative technique it may take no more than five minutes of microscope time to establish that it is worthless. On the other hand, a single sample of experimental material may take hours and days of time to evaluate and record. One day, all that is needed is a visual check that does not require particularly good performance on the part of the microscope; the next day, high resolution pictures may be required that demand the utmost performance from the instrument, and perhaps a half day will be spent "tuning" the microscope for this purpose. These problems work themselves out very efficiently in a small cooperative group without any formal scheduling. Flexibility is an asset that leads to productivity.

There is an inevitable tendency for electron microscope laboratories to grow by adding microscopes as well as personnel. Within limits this is highly commendable. In more than one instance that this writer knows of, however, quality has deteriorated as the quantity grew. Obviously there is danger of the director and his well trained associates becoming swamped by the demands of instrumentation and training the inexperienced. The really successful large laboratories are those which support more than one highly qualified investigator.

1.7. Educating Administrators and Clinicians

The director of any laboratory of electron microscopy doing tissue work will be subject to heavy pressures from administrators and clinicians who are apt to have some inkling of the "wonders" of electron microscopy but little knowledge of its problems. They think of electron microscopy as a simple extension of light microscopy, and assume that problems that can be straightforwardly dealt with in the latter instance, should be all the better answered by the new technique. One has to convince these people that electron microscopy inevitably proceeds at a much slower pace so that the "simple" problem which might take only weeks with a light microscope may take a similar number of years with the electron microscope. Furthermore, one must persuade these people that the sluggishness has nothing to do with one's energies or abilities, but is intrinsic to the method. In short, these people must be educated!

A few figures relating to dimensions help explain the problem. If one uses the standard "200 mesh" supporting screens, there is just about 1 mm^2 of open area that can be viewed by microscopes which take ⅛-in. screens. Even assuming that this square millimeter might be covered by perfect sections (which would be as rare as a Royal Flush in honest poker) it would take screens by the hundreds to be the equivalent of one ordinary section that the pathologist perhaps deals with in five minutes or less. When one has good and interesting material it is quite reasonable to spend a half day or more on a single grid. In this time one sees less than a square millimeter of surface, and cannot possibly photograph more than a tenth of this even at low magnification. At this rate it would take a solid year of work to record what the light microscopist can dismiss while he smokes one cigarette.

If we juggle the figures differently, and take a picture at only $\times 1500$ on a 2-in. plate, we record but 0.001 mm^2. It is a rare day that one has material that warrants taking more than 50 pictures (which, of course, would not generally be at such a low average magnification). Even so, though, it would take the better part of a month actually at the microscope to photograph 1 mm^2, and 7½ years of continuous photography to photograph 1 cm^2.

If one considers the third dimension, the problem is, of course, much worse. Useful sections can be no more than 1/20 μ thick. Twenty serial sections traverse a single micron. They perhaps span the thickness of one mitochondrion. It would take 100 sections to get through a nucleus, and 200 sections to record the volume of a single small cell. The conventional embryologist rarely needs as many sections to go through an entire pig embryo!

An electron microscopist who is concerned with tissue work simply cannot accept most problems that are primarily statistical. He cannot possibly process enough samples. He can only effectively deal with problems which yield simple qualitative answers. Processes that involve development and change are his anathema. He must design experiments so that a relatively few samples will be significant. It is not only the administrator and clinician who sometimes must be educated in this, but also the ambitious beginner in electron microscopy who must learn to limit his objectives and define them carefully.

Electron microscopy received widespread publicity in its early stages, and undoubtedly many people were naively led to expect too much. It has its limitations! In essence one must pay for magnification by accepting a restriction of field and all that this implies. One cannot do conventional histology with an electron microscope any more than one can do gross anatomy with a light microscope. The world of the electron microscope is a world that could only be studied inferentially before. It is a vastly important world and we should hardly have to defend its exploration with the most effective tool to date. The flea scurries about and his whole world is less than one square yard of a dog. The electron microscopist can be very busy too, and in his whole lifetime not cover as much territory.

2. Tissue Exposure

2.1. Introductory Remarks

Every piece of tissue that is finally examined in the electron microscope represents a substantial investment of time and nervous energy. One cannot afford to waste this investment on haphazardly preserved material. In the long run any amount of effort that may be expected to improve preservation must be regarded as an economy.

There are two aspects to quality tissue preservation. First, and obviously, is the fixative itself. Second, and often unappreciated, there is the problem of getting it to the tissue. It can be said that we have one excellent, general-purpose fixative for electron microscopy at the present time. This is suitably buffered osmium tetroxide which is discussed in Chapter 3, but it often requires considerable ingenuity on the part of the investigator to utilize this most effectively in a particular situation. Even the most experienced investigators expect difficulties as they approach some new tissue or organ. There remain today parts of the mammalian body that so far have defied good preservation and effective study by electron microscopy. For instance, no one yet has published a good picture of a large muscular or elastic artery although we can be sure many efforts have been made to accomplish this.

The problem is aggravated by our virtual dependence on osmium tetroxide as the fixative of choice. For many years it has been realized by conventional microscopists that this fixative penetrates poorly. If one expects to bring substantial concentrations of it into contact with still living cells the latter must be very near an exposed surface, in most instances within a half millimeter of the surface.

Palade (1952) in his already classic study of buffered osmium tetroxide keenly appreciated the importance of small size in tissue block preservation. In retrospect the way he minced tissue was as essential to his success as the formula of his fixative. The beauty of Sjöstrand's (1953) study of retinal rods, also a classic, depended in part upon the good exposure of these elements that naturally occurs when the sensory retina is pulled away from the pigment layer.

But in terms of tissue blocks, how small is small? This writer is reminded of chatting with a colleague at a national meeting when a novice inquired about the sizes of pieces of tissue that might be well preserved. My experienced friend asked him to visualize the head of a pin, and then commented that we divide that into six parts! Although this may be a bit of an exaggeration, the beginner has trouble grasping how truly tiny, blocks must be.

There are, of course, several approaches to the problem of getting fixative to tissue. These are discussed in the following sections.

2.2. Minced Tissue

There are obvious advantages to mincing fresh tissue directly in fixative. One can start with tissue from any place in an animal body, even deep within an organ, or from vital areas. With practice and skill the time taken to mince tissue can be reduced to a minute or even less from the time the circulation is interrupted in a narcotized animal. Even if an animal is decapitated it need not take appreciably longer to find and remove parts of most organs.

The most serious objection to mincing is the inevitable mechanical disruption of the tissue. This can be minimized by using only brand new razor blades for cutting. But whether or not this is an important factor depends most directly on the character of the tissue. Thus, small acini of glandular organs may be expected to survive quite intact. Although distortions would be expected to occur in the connective tissue stroma, a relatively homogeneous cellular tissue such as liver minces easily. Tissue of the central nervous system, unsupported by connective tissue, usually is very difficult to cut finely in a satisfactory manner. Lung collapses under these circumstances, and testis falls to pieces, although segments of tubules are well preserved.

Probably the simplest and best way of mincing tissue in osmium tetroxide fixatives is to place a drop or two of fixative on a piece of cardboard with a relatively nonabsorbent surface (ordinary filing cards serve very well), or upon a thin sheet of dental wax. A small slab of tissue is excised from the animal as quickly as possible, and transferred into the drop of fixative with a toothpick. This slab or ribbon of tissue should be as thin as possible, certainly less than a millimeter thick, and perhaps 2 or at the most 3 mm long. The toothpick then serves to pin the tissue down while the tissue is minced with a razor blade (Fig. 1). The latter must previously have been cleaned of the protective oils which the manufacturer puts upon it.

The actual mincing must proceed with the utmost speed and be stopped within 15 or 20 sec even though not all of the tissue may be finished. It is possible later to select only those pieces of satisfactory size so it is not important at this stage to have uniformly small tissue blocks. One reason for halting the mincing quickly is that osmium tetroxide reacts with the steel of the razor blade so that a black precipitate appears. No doubt it is advisable to stop as soon as this becomes evident. But even more important seems to be that tissue fluids and plasma destroys within seconds the desirable properties of buffered osmium tetroxide fixative unless these substances are greatly diluted. Thus, the single drop of fixative in which one minces will very quickly do more harm than good. It is therefore essential to get the tissue into a proportionately large volume of fixative within seconds.

The transfer of the tissue to a container with a large volume of fixative can be done most easily if the original mincing was done near the tip of a fairly narrow strip of cardboard, a strip which will fit into the vial of fixative (Fig. 1). Ordinarily most of the tissue pieces will simply float away when the tip of the cardboard is immersed in the vial. Any pieces that remain adherent to the cardboard can be flicked off with a toothpick. Strips of dental wax may be substituted for cardboard, hard or soft grades are obtainable from any dental supply house. In this procedure, metal does not come in contact with the main volume of the fixative at all. If for any reason tissue must be picked up and trans-

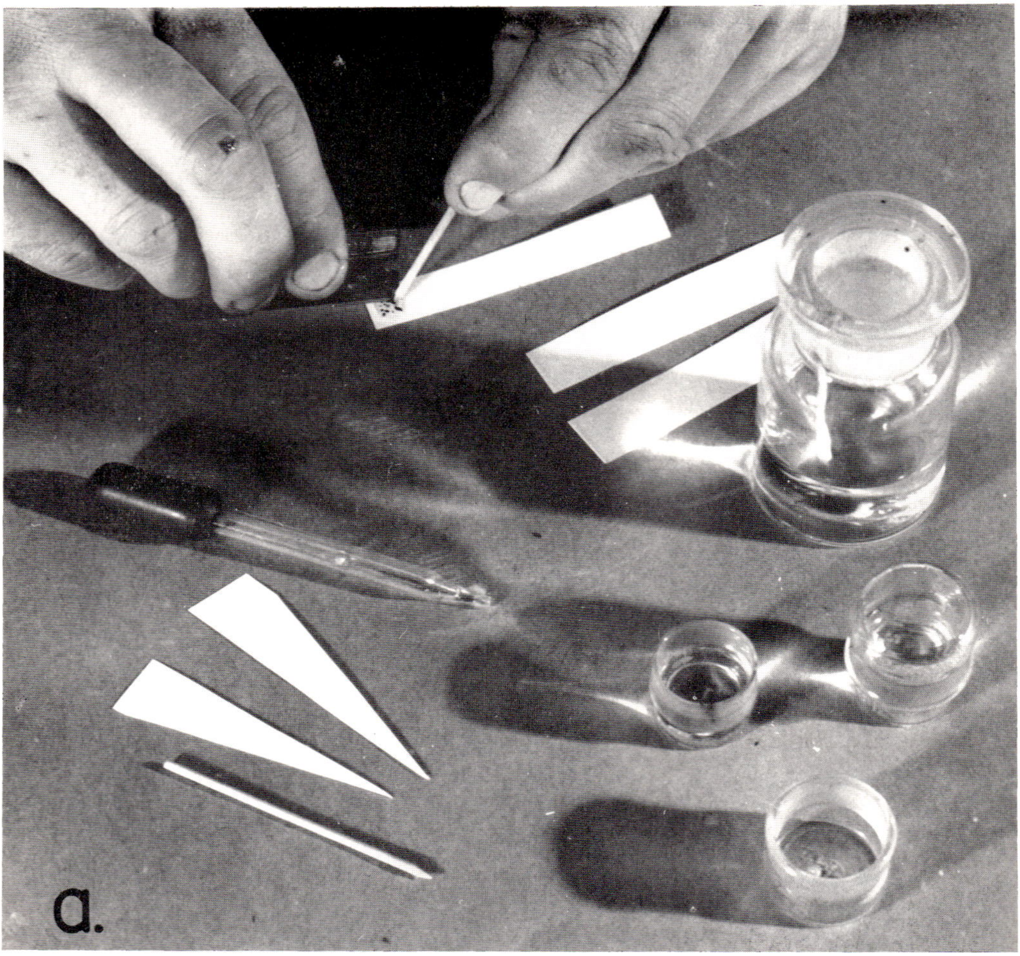
a.

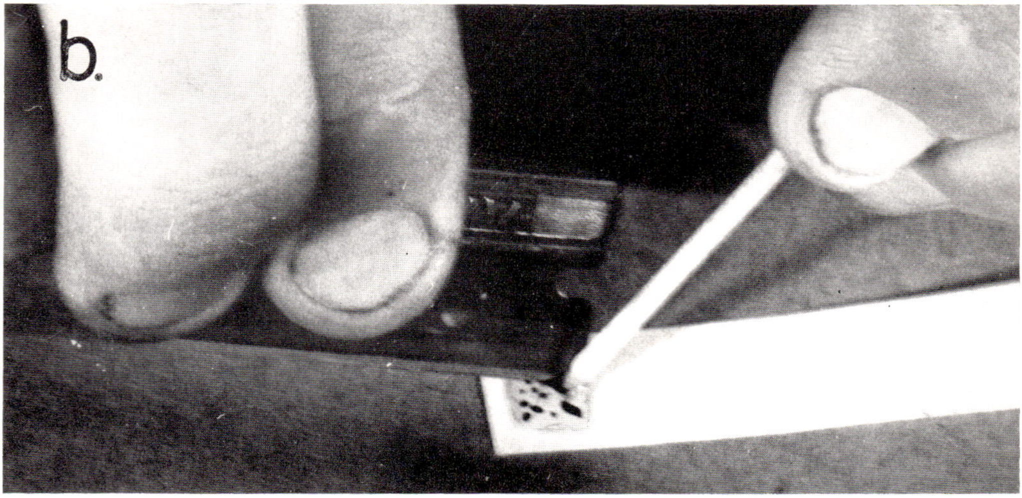
b.

ferred to the vial of fixative, toothpicks or cardboard points (Fig. 1) should be used rather than needles or forceps. If appreciable quantities of plasma or tissue fluids are transferred to the vial, the first fixative should be discarded at once and replaced with fresh. It probably helps, too, to agitate the vial for the first 5–10 min.

Since blood and tissue fluid quickly destroy osmium tetroxide fixatives, any large excess that may be picked up with a fresh piece of tissue should be blotted on a piece of filter paper before the tissue is put in fixative and minced. Do *not* make an attempt to wash tissue with so-called physiologically balanced salt solutions, at least unless careful controls are studied, for almost invariably they do demonstrable damage.

There are occasions (such as when using permanganate fixatives) when the tissue must be ice cold when it is in the fixative. The tissue almost has to be minced "dry," in its own juice, on a strip of cardboard, and subsequently transferred to a previously chilled fixative. Obviously speed is particularly important under these circumstances, and one should be careful not to have an absorbent cardboard.

In working with osmium tetroxide fixative the operator must appreciate the fact that its fumes are dangerous. They are quite capable of fixing corneal epithelium. Operations should be done by an open window or in the cross-current of a fan, or even in a chemical hood. The latter is not really necessary for fortunately osmium tetroxide fumes warn of their presence by their characteristic, powerful smell. If one does not hold his breath he is not apt to move his head in too closely. Some people will wish to wear

FIG. 1 (a) and (b). Mincing tissue in a drop of osmium tetroxide on a strip of cardboard. The operator must work very fast and not attempt to sort out pieces of correct size until later. The specimens should be transferred to a glass container with a surplus of fixative in it within a minute. This can be done most simply by dipping the tip of the cardboard strip in the bath, and using a toothpick or cardboard point to push the specimens. Metal should be avoided as far as possible, although a new clean razor blade must be used for cutting. This is discarded when black deposits form on its corners. Pieces of tissue are maneuvered and held in place with toothpicks rather than metal needles. Small, wide-mouthed, "weighing bottles" with ground glass lids of the type indicated here are almost ideal containers for the minute specimens during fixation and dehydration.

gas tight goggles for this work which, no doubt, is a good precaution. Most people will also find that a low power binocular viewer attached to a head loop will be advantageous. It is quite possible to wear such a loop over a suitably chosen pair of goggles.

2.3. Topical, *in Situ*, Preservation of Natural Surfaces

It is often possible to take advantage of naturally occurring surfaces to preserve tissue in a living animal while blood continues to circulate through the tissue in question. Under these circumstances mechanical damage can be kept to an absolute minimum. It is a particularly appropriate means of preservation in an organ like the kidney which receives its blood supply from within. It is less appropriate in anatomical situations like the adrenal gland which receives blood by way of its capsule, and thus would lose its blood supply before fixation begins. It has proved an important way of preserving tissues which are too delicate to withstand well the mechanical disturbance of mincing, such as those of the central nervous system. It is also a good way to fix muscle in a normally extended state.

If an organ surface is to be fixed *in situ,* it has been found almost essential that its capsule be removed. In general, enclosing connective tissue membranes prove to be very effective barriers to osmium tetroxide penetration, and one cannot get adequate fixation through them. This is true even when the capsule appears to be very delicate in a gross sense. Thus, the capsule of a mouse kidney or the pia of a rat brain, delicate as these structures seem, demonstrably block the penetration of fixative. These membranes, however, can be peeled off without difficulty by grasping them with finely sharpened jeweler's forceps and tearing. Fixative is then flushed on the exposed surface.

It is of the utmost importance that any oozing blood or tissue fluid be flushed away quickly with excess fixative. The fixative, of course, will soon seal the tissue so that oozing will stop. This flushing can be done most simply with the medicine dropper type of pipette. A few squirts and gentle rubbing with the blunt tip of the pipette are all that are normally needed. Do not, however, take a full pipette and squirt and suck. This contaminates all of the fixative. It is better to begin with 3 or 4 drops, and to keep repeatedly flushing these away until the exposed area is cleaned. One must be particularly careful not to contaminate the main bottle of fixative with the tissue exudates.

Before beginning these procedures the animal should have been so arranged or tilted that excess fixative can drain away from the pertinent area. It is often expedient to let it drain into cotton pads. Osmium tetroxide fumes escape in dangerous quantities. A person should be most careful of his eyes while doing this. It is advisable to have a small fan blowing the fumes away from the operator. One must also breathe naturally for the nose gives warnings of concentrations dangerous to the eyes. The danger lies in holding the breath and carelessly moving in close.

Ordinarily a freshly exposed organ surface can be sealed and cleaned within 2 or 3 min. From that point on, one need only trickle fixative through a thin pad of absorbent cotton to continue fixation. The thinnest wisp of cotton is placed on the organ surface, just enough to keep a fluid layer on the surface so that it does not require too much attention. But the pad should be so thin that the addition of 2 or 3 drops of fixative will amount to a fresh bath. The procedure then is to add a couple of drops of fixative to the cotton every 2 or 3 min.

After 10 or 15 min it can be assumed that the fixative has penetrated as deeply as it usefully will. At this time the preserved surface should be undercut and the thin slab of tissue so obtained transferred to a vial of fixative. The fixed tissue, of course, has been hardened so that ordinarily the undercutting can be accomplished fairly exactly at the boundary between fixed and unfixed tissue. If any large amount of unfixed tissue is removed it sometimes can be scraped or cut away while the tissue is held on the tip of one's finger, and before putting it into the vial of fixative, or it can be removed later during dehydration.

The interior linings of organs may present special problems in the preservation of their natural surfaces. It is often tempting to try a perfusion of fixative, either through a cannula, or simply by direct injection from a syringe. This may be highly successful if the lumen is small or of narrow bore so that the organ can be flushed rapidly and adequately with a great surplus of fixative. To accomplish this, a drainage path must be established some distance away from the site of perfusion. Although some organ linings may withstand washing with physiologically balanced salt solutions, the procedure is dangerous under the best of circumstances and its effects should be carefully controlled. It is obviously impractical to flush a large organ with anything but a saline solution. But when organs are of such size, one can open them surgically and achieve a topical, *in situ* preservation as discussed previously in this section. The contents of some organs, such as the gall bladder and parts of the gut, tend to destroy the proper action of the fixative to such an extent that they require particularly vigorous flushing.

2.4. *In Situ* Fixation of Cut Surfaces

One often has the problem of wishing to study the deep part of an organ. Sometimes this can be done best by *in situ* fixation of a surface artificially created by simply slashing across the organ. Fixative is then applied as described in the last section. This method is applicable also to organs which do not have capsules that can be mechanically removed, for example, salivary glands, lymph nodes, marrow, etc. Also, when the afferent of blood supply is by way of the capsule and would have to be sacrificed in any case, (such as the adrenal gland), the method would seem to be advantageous. The technique often is preferable to mincing tissue, in that delicate stroma is not so severely damaged.

There are apt to be more or less severe difficulties with bleeding when an organ is transected. This, of course, will depend upon the organ, but much more flushing with fixative may be required to seal the surface than if a natural surface is used. Cotton swabs may be used to control the first more or less massive bleeding, followed by vigorous flushing with fixative. Avoid the use of salt solutions for this purpose. In general, success with this method will depend upon anatomically avoiding the larger blood vessels so that massive hemorrhages do not occur.

Once again it is pertinent to stress that the eyes must be protected from the fumes of osmium tetroxide, for under the pressure of working efficiently and fast, it is all too easy to forget about one's eyes until corneal damage has been done.

2.5. Vascular Perfusion

It is natural for one acquainted with the techniques of conventional microscopy to think of vascular perfusion as a likely way of preserving tissue for electron microscopy. Yet, as far as osmium tetroxide is concerned, this is doomed to failure in all but very special cases. There are at least two reasons for failure. Decisive under most circumstances is the fact that osmium tetroxide acts as an extreme vasoconstrictor so that the arteriolar bed is clamped shut as the first fixative reaches it. Little or no fixative gets beyond this point in an arterial injection. This is true even though efforts are made to counteract the vasoconstriction by procedures that can be expected to dilate. Thus, in the author's own laboratory, we have perfused with vasodilators before running in the fixative. We have used gaseous vasodilators such as amyl nitrite. We have precooled animals so that their blood temperature was close to zero before making the perfusion. And we have killed animals by building up blood carbon dioxide levels close to the lethal point before beginning perfusion. None of these methods made possible a satisfactory general perfusion. It is reasonable to think that similar experiments must have been attempted in numerous other laboratories and these, too, must have failed, for the literature does not report success with this method.

There are limited anatomical situations where vascular routes can be used for the perfusion of fixative. Perhaps most promising are those in which a retrograde injection through venous channels is reasonable. Liver and spleen are obvious examples. Excellent micrographs of liver cells preserved by vascular perfusion have, in fact, been published. Of course, other body cavities, such as the cerebrospinal fluid passageways, can be used as routes for the *in situ* preservation of neighboring tissues.

When natural body channels are to be used for the dissemination of fixative it is important to realize that a large excess of fixative must be pumped through rapidly. That is to say a drainage route must be opened up so that flushing can be vigorous at first. This is necessary since the desirable qualities of the fixative are quickly destroyed upon mixing with body fluids and upon contact with tissue. No doubt this is partly due to the removal of osmium tetroxide by combination with organic material. Probably equally important is the loss of buffering capacity and the ionic imbalance which results quickly under these circumstances. In any case it is quite inadequate to get small quantities of fixative to a region, perhaps, as is the usual result, just enough to blacken the tissue spottily.

2.6. Hyaluronidase Pretreatment

Dense connective tissue presents a formidable barrier to the penetration of osmium tetroxide fixatives. This is true whether one is considering an organ capsule, an adventitial sheath, or simply an irregularly arranged tissue such as dermis.

When such tissues are to be fixed, they first may be exposed to a brief treatment with strong hyaluronidase which presumably acts by depolymerizing the mucopolysaccharides. There is good reason to believe that subsequently the fixative penetrates much more effectively (Pallie and Pease, 1958).

The method which has been developed consists in using a commercial preparation of lyophilized hyaluronidase available in 150 unit ampules (Wydase, manufactured by Wyeth Laboratories Incorporated, Philadelphia, Pennsylvania). This is dissolved in 2–10 ml of normal saline or Ringer's solution. It is either topically applied *in situ* to a surface to be fixed, or small pieces or thin sheets of tissue are submerged in the solution. Reaction is complete within 5 min, so that longer exposures should not be employed.

Fixative follows this treatment without intervening washing of any sort. Three precautions seem necessary. The enzyme is effective at room or body temperature, but not when iced. The enzyme is quickly destroyed by body fluids, so that the surface to be prepared must be flushed with an excess of the solution. In particular, any oozing blood must be flushed away. In the case of excised tissue, the first bath is changed quickly to a second and even a third bath, and agitation is maintained. Salt solutions should not be used for preliminary flushing or washing although oozing blood or tissue fluid can be blotted away with absorbent cotton.

The initial investigation of the use of hyaluronidase was unable to detect any deleterious effect upon the subsequent preservation of tissue. The method has not had wide application as yet, however, and so should be used cautiously in new investigations. It is advised only when a serious connective tissue barrier must be eliminated.

2.7. Special Problems with Human Tissues

The previous sections of this chapter have emphasized ways of bringing tissue in contact with fixative in the most expeditious manner. Obviously most of these procedures are not generally applicable when human material is to be obtained. Anyone acquainted with the literature of tissue studies by electron microscopy will realize that published micrographs of human material rarely equal in quality the best that can be obtained from animals. The difficulties in getting well preserved human material very seriously retard the development of pathological knowledge at electron microscopic levels. Since these problems are largely unsolved, this writer can do little more than define their nature and indicate hopeful approaches.

Needle biopsy specimens usually have proven disappointing, although even experienced investigators report that one may prove poor while the next can be fairly good for completely mysterious reasons. No doubt speed is a limiting factor as well as mechanical disruption. In addition it should be realized that needle biopsies are apt to contain much tissue fluid and tissue debris. As indicated previously this can be expected to destroy the properties of the fixative quickly. No doubt one should pass the tissue quickly through one or two changes of fresh fixative to flush away this material within the first few seconds. In this connection blood samples and sternal puncture marrow does not preserve as well as one would like because of the inevitably associated serum.

This writer knows of several pathologists, currently doing some of the most beautiful work with human material, who have completely given up needle biopsies, and insist upon surgical biopsies instead. Even the collection of these under the best circumstances presents problems. The surgeon quite naturally likes to block off circulation before removing tissue. Yet this is apt to do irreparable damage from our point of view. Furthermore, the surgeon must be educated not to maul the tissue or leave it sitting on towels. The responsible investigator, rather than a technician, had better be in the surgery himself, and must have discussed the problem with the surgeon in advance. Then a small snip of otherwise undamaged tissue, promptly minced, may yield satisfactory results.

One might think that satisfactory surgical biopsies of the skin or of the mucous membranes of body orifices would be easy to obtain, yet experience has shown that there are great difficulties, and very few good micrographs have been published. It seems reasonable to think that a tissue infiltrated with a local anesthetic will not be satisfactory, so emphasis should be upon tissue obtained under general anesthesia. These regions of the body, in general, have a dense connective tissue dermis or lamina propria. Fixation might be improved with hyaluronidase pretreatment as discussed in the previous section.

It is common knowledge and experience that tissues of young animals can be preserved well more easily than those of old adults. The reasons for this are not altogether clear, but in part, may be due to differences in the amount and character of connective tissue elements. One can expect these differences to be present also in dealing with human material, and, indeed, every pathologist recognizes them to a degree at the level of conventional microscopy. Where possible, therefore, efforts should be made toward obtaining juvenile or young adult tissue.

The human, after all, is a large animal with consequently much more and much denser connective tissue in most places than a mouse or a rat. Thus, tissue which preserves easily when taken from a small animal may present serious problems when removed from humans. A prefixation treatment with hyaluronidase may prove to be more generally necessary.

Although this section may seem discouraging to those who would do human pathology, this field promises particularly rich rewards, and ways to success in particular situations are being found and will be found. One should not be unduly discouraged if one's first efforts are dismally disappointing. It is certainly advisable for the novice to have a taste of success first by studying the normal morphology of the organ in question in a small laboratory animal before plunging into human pathology. In doing this he will learn what is potentially possible. It is only when a person has gained perspective in this way that he is really in a position to appreciate his problems and his degree of accomplishment.

2.8. Special Problems with Tissue Cultures and Cell Suspensions

One might think that tissue cultures with their exposed cells would be the easiest of all tissues to preserve well, but this has not proven to be the case. Some of the difficulties, no doubt, relate to the properties of the fixative itself, and these are discussed in the next chapter. "Polymerization damage" is particularly prone to affect exposed cells when conventional methacrylate embedding is employed. This, and techniques for eliminating it, are discussed in Chapter 4, but it should be pointed out here that tissue culture media no doubt behave rather like serum and tissue fluids. As has previously been discussed, these act powerfully to destroy the desirable qualities of osmium tetroxide fixatives so that vigorous flushing is necessary.

Thin sheets of tissue culture cells may prove difficult to see when it comes time to orient blocks for sectioning. They can be stained with eosin or, better, a basophilic dye such as toluidine blue while they are being dehydrated which will facilitate subsequent handling.

In the case of cell suspensions, extreme dilution in a very large volume of fixative with agitation may improve preservation. However, it has several times been reported better to centrifuge suspended cells into a compact mass and then to treat small aggregates as though they were solid tissue blocks. Thus, the "buffy coat" of centrifuged blood can be handled effectively in this way (Low and Freeman, 1958), as also can be particulate fractions of cells (Palade and Siekovitz, 1956).

In dealing with blood in particular, it is desirable to avoid coagulation. Low and Freeman (1958) advocate the use of silicone coated glassware. This is prepared by its immersion in a 2% solution of Dow–Corning #200 antiwetting agent dissolved in carbon tetrachloride, followed by baking for 0.5–1 hr at 450–550° C. Hypodermic needles and/or other steel surfaces are coated with Armour Company "Monocote" by immersion, followed by draining and air drying.

2.9. The Use of Cold-Blooded and Young Animals

People using tissues from cold-blooded animals have a great advantage over those who insist upon studying mammals. These tissues apparently can stand much more abuse. This is particularly true if the animals have been prechilled so that the tissues are cold. Dissection then can be a relatively leisurely affair, and the animals can even be dead for perhaps as much as a half an hour and still yield well preserved material.

To a certain extent newborn mammals can be treated similarly. Usually tissue from such babies can be removed after decapitation and still be well preserved in spite of the inevitable delay. This is certainly simpler than trying to keep a very young animal alive under anesthesia while a delicate operation is performed. If the removal of tissue is to be at all complicated, and take unduly long, the animal can be precooled before decapitation.

2.10. Botanical Material

The fact that plant material is likely to be difficult to preserve has led people to experiment with variations of the fixatives, some of which are discussed in the next chapter. McLean (1958), however, emphasizes that the presence of large turgid vacuoles creates a problem not found in animal cells. He supposes that the destruction of the differential permeability of the tonoplast as the fixative penetrates may result in the sudden release of acidic vacuolar fluid into the as yet incompletely fixed cytoplasm. He proposes that the turgor pressure be released before fixation by treating the plant tissue with suitable glucose solutions until plasmolysis is incipient. This has clearly proved effective in some instances.

Literature Cited

Low, F. N., and Freeman, J. A. (1958). "Electron Microscopic Atlas of Normal and Leukemic Human Blood." McGraw-Hill, New York.

McLean, J. D. (1958). *Program (Abstr.) Fourth Intern. Conf. Electron Microscopy, Berlin,* p. 81.

Palade, G. E. (1952). *J. Exptl. Med.* **95,** 285.

Palade, G. E., and Siekovitz, P. (1956). *J. Biophys. Biochem. Cytol.* **2,** 171.

Pallie, W., and Pease, D. C. (1958). *J. Ultrastructure Research* **2,** 1.

Sjöstrand, F. S. (1953). *J. Cellular Comp. Physiol.* **42,** 15 and 45.

3. Fixation

3.1. Introductory Remarks

There is no question but that osmium tetroxide fixatives are the best that we know of today. It is historically interesting that these were used from the very first by electron microscopists attempting to work with tissues. There was, of course, some reason for this since discerning cytologists long had recognized osmium tetroxide as a superior preservative of cytological detail, and it was common practice to use its vapors to fix delicately protozoa and tissue cultures. As electron microscopes became generally available, after the Second World War, several laboratories made efforts to grow tissue culture cells in such a flattened manner that they could be used as whole mounts in an electron microscope (cf. Porter (1953) for early references). Osmium tetroxide was quite logically their fixative of choice.

During this period a few ultrathin sections of tissue were obtained by somewhat heroic means. Claude and Fullam (1946) published the most revealing micrographs of sectioned material to that date, and these had been preserved with osmium tetroxide. Pease and Baker were able to cut substantial numbers of moderately thin sections in 1948, and followed the lead of Claude and Fullam in using osmium tetroxide. Before two more years passed, three different laboratories had completed comparative studies of osmium tetroxide, and a variety of other more or less common fixatives. All three laboratories were in common agreement that osmium tetroxide was clearly the fixative of choice (Dalton, *et al.*, 1950; Porter, 1950; Pease and Baker, 1950). Subsequent work has amply confirmed this decision.

In those early days, of course, it was not clear just what quality might be obtained in thin sections of preserved material; and, in fact, osmium tetroxide often seemed to behave quite erratically. It remained for Palade in 1952 to demonstrate what might be done. He developed a slightly alkaline, buffered osmium tetroxide solution that gave reasonably consistent results, and which was clearly superior to acidic fixatives or unbuffered osmium tetroxide (which produces an acidic wave of injury as it penetrates). This is a solution which has since come to be known commonly as "Palade pickle," and which is used by most laboratories with little or no modification of the original formula.

The following year F. S. Sjöstrand brought from Sweden to the national meeting of the Electron Microscope Society of America the most impressive collection of electron micrographs of ultrathin sections that had been seen. At that time Sjöstrand was using Palade's fixative, and obviously he applied it with considerable skill. It should be added, though, that the excellence of his demonstration depended not only upon preservation but also upon his superior sectioning and skilled microscopy. The net effect was to set standards which, from then on, had to be met by every laboratory to consider itself successful in this field.

3.2. Palade (1952) Fixative, Buffered Osmium Tetroxide

There is an extraordinary unanimity amongst experienced electron microscopists that osmium tetroxide fixatives are always the general purpose fixatives of choice for all tissue work. The beginner should by all means successfully accomplish what is possible by these means before turning to other preservatives. He will not be seriously handicapping himself if he begins with the original formula proposed by Palade in 1952. This simply uses a rather strong buffering solution with osmium tetroxide added. The buffer chosen was the acetate-veronal buffer of Michaelis. Apparently Palade empirically found this to be superior to inorganic buffers, and there was the further theoretical consideration that the bivalent ions of tissues would not form insoluble precipitates with it.

The fixative is prepared as follows:

Stock buffer solution:

Sodium veronal (barbital)	14.7 gm
Sodium acetate	9.7 gm
Distilled water to make	500 ml

Stock osmium tetroxide solution:		
Osmium tetroxide*	2%	
Fixative:		
Buffer	5	ml
0.1 *N* HCl†	5	ml
Distilled water	2.5	ml
2% osmium tetroxide solution	12.5	ml

The fixative should have a pH of about 7.4. Small corrections are made by a few additional drops of buffer or acid as required before the distilled water is added to make the 25 ml.

Both the buffer and the fixative can be kept a reasonable length of time in a refrigerator. At room temperature micro-organisms may grow in the buffer. On the whole it seems safest to make up relatively small quantities of buffer and fixative at one time and to use these up within a few weeks.

Any person with the least acquaintance of tissue studies by electron microscopy has seen numbers of electron micrographs of cells preserved with the Palade fixative, or one of its close variants. In this book, all of the tissues, unless otherwise noted, were of material prepared in this way. Figure 2 is a representative sample.

3.3. Variations of Osmium Tetroxide Fixatives

The variations of Palade fixative that have been proposed include alterations in its strength, its molarity, its pH, and the

* Special precaution must be observed in preparing osmium tetroxide solutions. The operator's respiratory tract, and particularly his eyes, must be protected from the fumes when the ampule, in which it is packaged, is broken. It is best to smash this under water in the container which is to be used for storage.

It is also essential that all glassware be free of organic contaminants for these will combine with, and destroy, the osmium tetroxide. For this reason only glass-stoppered bottles should be used. And for this reason, before breaking the ampule, its label must be soaked off, and all traces of glue removed.

Osmium tetroxide dissolves very slowly, so that solutions containing it must be made up at least one day before their intended use.

† One can make a 0.1 *N* HCl solution by diluting 8.6 ml of concentrated HCl (36%, 11.6 *M*) in water to make 1 liter.

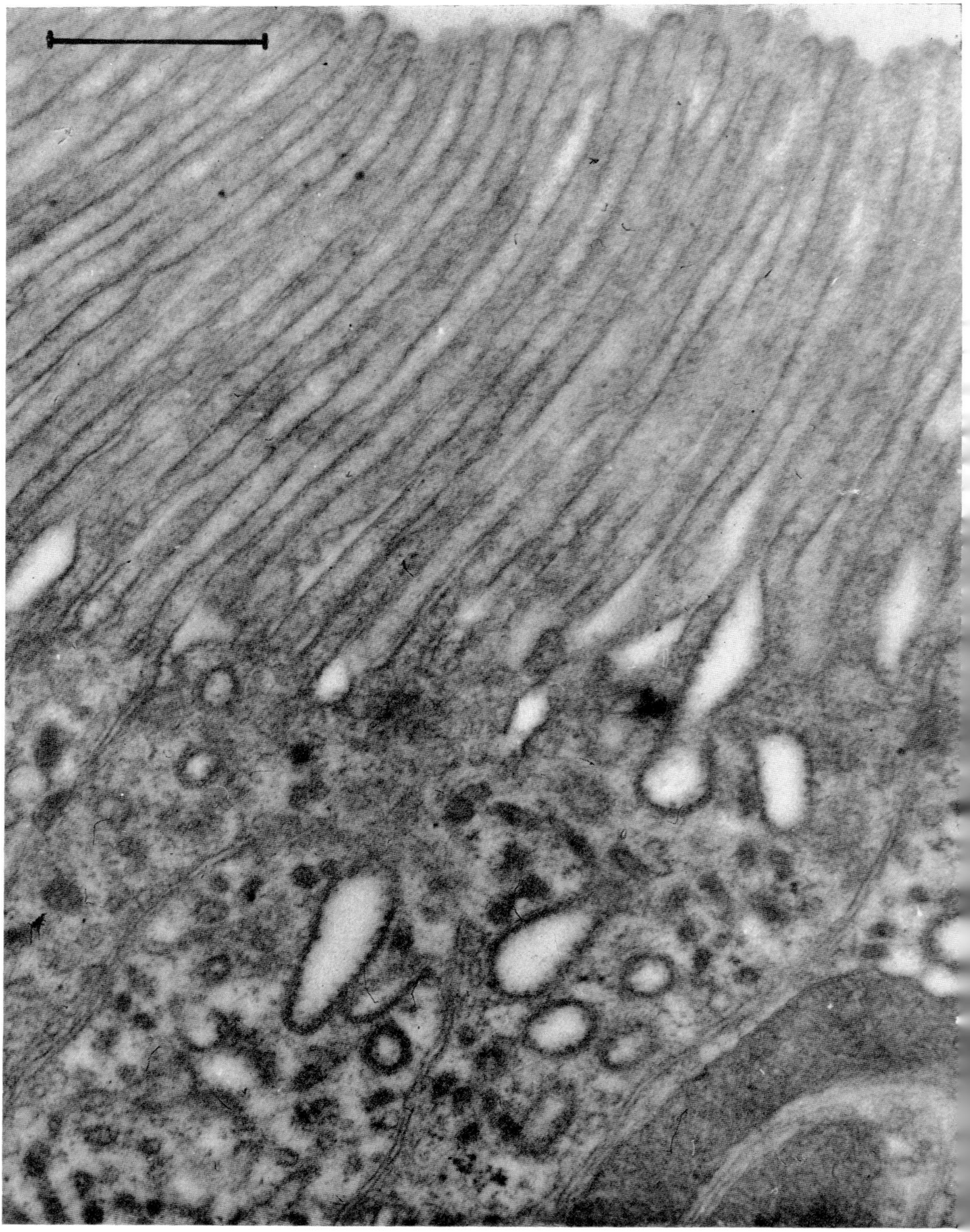

effects of particular ions. Palade, giving a paper at the Third International Conference on Electron Microscopy in 1954 (published in 1956), was able to reassess fixation problems in the light of much experience, and he discussed some of these.

Insofar as strength is concerned, it may be mentioned that Palade originally suggested that a 2% variation of his formulation might be used, and later Palay and Palade (1955) suggested that even 3 and 4% mixtures gave better results with tissues "particularly rich in reducing substances," in this instance tissue of the central nervous system. The present writer attempted to make careful comparisons of 1 and 2% solutions when he was studying the kidney by *in situ* preservation. The stronger solution seemed definitely preferable, and he has mainly used it since. Perhaps it can be said that the added strength is sometimes necessary, but more often wasteful.

The group at the Rockefeller Institute recognized almost from the beginning that the Palade fixative was hypotonic, but they discounted this as of little importance. However, Caulfield (1957) has published a variation in which a good deal of sugar is used to increase tonicity. Since this was published as coming from the Rockefeller Institute, it may be considered the heir apparent to the original formula. One starts with the standard 1% buffered osmium tetroxide fixative, and adds 0.045 gm of sucrose per ml of solution.

Caulfield has recognized that there are special problems associated with the preservation of *botanical material* and for this recommends a fixative with much less salt, as follows:

Stock buffer	1 part
0.1 *N* HCl	1 part
Distilled water	2 parts

FIG. 2. The apical region of a cell of the proximal tubule of a rat kidney. This material was fixed in buffered osmium tetroxide, blocked in butylmethacrylate, sectioned and mounted on a single supporting film, and lightly stained with phosphotungstic acid. Except for the last, these are basic and routine procedures which are commonly employed by most laboratories. This figure should be compared with sections mounted differently (Figs. 21 and 31a), embedded differently (Figs. 9 and 10), and fixed differently (Figs. 4, 5, and 6).

2% osmium tetroxide solution	4 parts
Sucrose	0.015 gm/ml

The fixative should have a pH of about 7.4.

People associated at one time or another with the Rockefeller Institute group have also experimented with the pH of the basic Palade formulation. These investigators generally agree that when the tissues are highly hydrated it is desirable to buffer way over on the alkaline side at a pH of 8, or even 8.5. In general, invertebrate and embryonic tissues, as well as protozoans, seem to be preserved better under these conditions.

People associated with the Rockefeller Institute group also have concerned themselves to some extent with specific ion effects. They have found, for instance, that the ammonium ion must be avoided in buffers for it destroys the membranes of the endoplasmic reticulum. On the other hand, calcium and/or magnesium ions in little more than trace quantities may be advantageous in preventing the extraction of "cement substances." This is, perhaps, particularly important with delicate and highly hydrated tissues (Palade, 1956).

Investigators at the Karolinska Institutet, led by F. S. Sjöstrand, have not felt in recent years that the hydrogen ion strength was the most important part of the fixation vehicle. Rather they have emphasized tonicity. Investigation of this was begun by Rhodin (1954) and culminated in the following formulation of Zetterqvist (1956). This mixture is basically the Palade fixative with balanced salts added.

Stock buffer solution:	
Sodium veronal (barbital)	14.7 gm
Sodium acetate	9.7 gm
Distilled water to make	500 ml
Stock salt solution:	
Sodium chloride	40 gm
Potassium chloride	2 gm
Calcium chloride	1 gm
Distilled water to make	500 ml
Fixative:	
Buffer solution	10 ml

Salt solution	3.4 ml
0.1 *N* HCl	11 ml
Distilled water to make	50 ml
Osmium tetroxide (dry)	0.5 gm

The fixative should have a pH of 7.2–7.4, and its tonicity is estimated to be 0.34*M*.

No adequate comparison of the Caulfield and Zetterqvist fixatives has been made. There probably would be no detectable difference in the deeper cells of most tissue blocks. Cells exposed at a surface, however, experience the effect of an osmotic imbalance with extreme rapidity, and are particularly sensitive to damage. Investigators who have most successfully dealt with tissue culture cells usually have added salts to the Palade fixative, and thus have used formulas much like the Zetterqvist one.

If the beginner is bewildered by this discussion he should rest assured that any one of the above formulations is quite capable of giving satisfactory results in most situations. If his first efforts are disappointing he should not blame the fixative, or jump to a new formula; rather he should consider the problems of reaching the tissue with fixative as discussed in the previous chapter. The real difficulty in obtaining good material for electron microscopy is almost always a difficulty of getting the fixative to the tissue under optimal conditions.

3.4. Temperature Control of Osmium Tetroxide Fixation

There are theoretical reasons for thinking that chilling osmium tetroxide fixatives might produce a more delicate preservation of tissues than would otherwise be obtained. At least one might expect that cytolytic activity would be minimized. A number of laboratories including those of the Rockefeller and Karolinska Institutes have at times advocated precooling the fixative to near 0° C temperatures. Palade has indicated his belief that thicker masses of tissue can be well preserved at low temperatures than would otherwise be possible. Rhodin has thought that there were qualitative differences as well as quantitative ones. The present writer confirms these opinions.

In our own laboratory we have on occasion compared tissue preserved at near 0° C with samples fixed at 21° C, 37° C, and 50° C. There were subtle qualitative differences among the four samples, although we did not push the investigation far enough to analyze these in detail. Clearly the higher temperatures should be avoided. We have the subjective impression that the best results follow a short initial fixation at 0° C, after which the vial with its tissue is removed from the ice bath and allowed to equilibrate at room temperature where it remains until finished.

In talking of low temperature fixation one must bear in mind that no doubt osmium tetroxide begins to act within seconds after the tissue is immersed, so that if tissue is to be minced in a drop of fixative, even this must be precooled, or the tissue may be minced "dry" and subsequently added to chilled fixative. If a tissue is to be fixed *in situ,* relatively large quantities of chilled fixative must be flushed over the surface to cool the tissue as rapidly as possible. Even then its effectiveness is certainly limited.

3.5. Duration of Osmium Tetroxide Fixation

In the early years of tissue work with osmium tetroxide solutions it was common practice to fix for relatively long periods of time, perhaps overnight or even longer. After 1952, however, it became apparent that not more than a very few hours gave much superior results in most instances. At the Third International Conference on Electron Microscopy in London in 1954 (published, 1956) Palade clearly and dramatically demonstrated that prolonged exposure to the fixative allowed proteinaceous substances to leach out of muscle tissue. Membranous structures, predominantly lipid in nature, were less affected than purely proteinaceous material. This suggests the deliberate use of long fixation to differentiate membranous structures under conditions when these tend to be masked. Ordinarily, however, it is obvious that a short fixation period is to be preferred when the "total" contents of a cell are to be preserved. It has since become evident that 30–60 min ordinarily is ample time for optimal preservation, and certainly one should not exceed 2 hr except under special circumstances.

In this writer's laboratory an effort was made to determine the minimum time that might yield successful preservation. Exposure for as little as 3 min produced as satisfactory fixation of kidney as any longer period, although, perhaps the depth of fixation was not as great. It must be realized, therefore, that "complete" fixation occurs very quickly, within minutes or even seconds of the time when the fixative reaches the tissue. Little or nothing is gained by prolonging exposure excessively, and deleterious processes start.

3.6. Criteria of Good Fixation

Criteria of good fixation perforce relate to the best that can be accomplished with osmium tetroxide. Admittedly there may be some organelles and/or inclusions that are not preserved adequately in this way. The most conspicuous deficiency is in the preservation of chromosomes, but there seems to be a general rule that almost anything that is reasonably well preserved by any chemical fixative, can be preserved as well or better by osmium tetroxide.

Different cell types differ so much, both qualitatively and quantitatively, that it is difficult to categorize the criteria of good fixation. However, the following generalizations may be considered guideposts. Figures 3(a) and (c) compare fair and bad fixation in one cell type, and illustrate most of the features discussed below.

Nuclei which are uniformly and finely granular are usually associated with other desirable features. Obviously aggregated chromatin masses are suspect; although, in a few cell types (i.e., plasma cells), this pattern is so dominant that it may almost always be evident.

Mitochondria ordinarily are fairly sensitive indicators of fixation. In many cell types they easily become obviously swollen and empty looking. Other cell features may be none the less fairly well preserved.

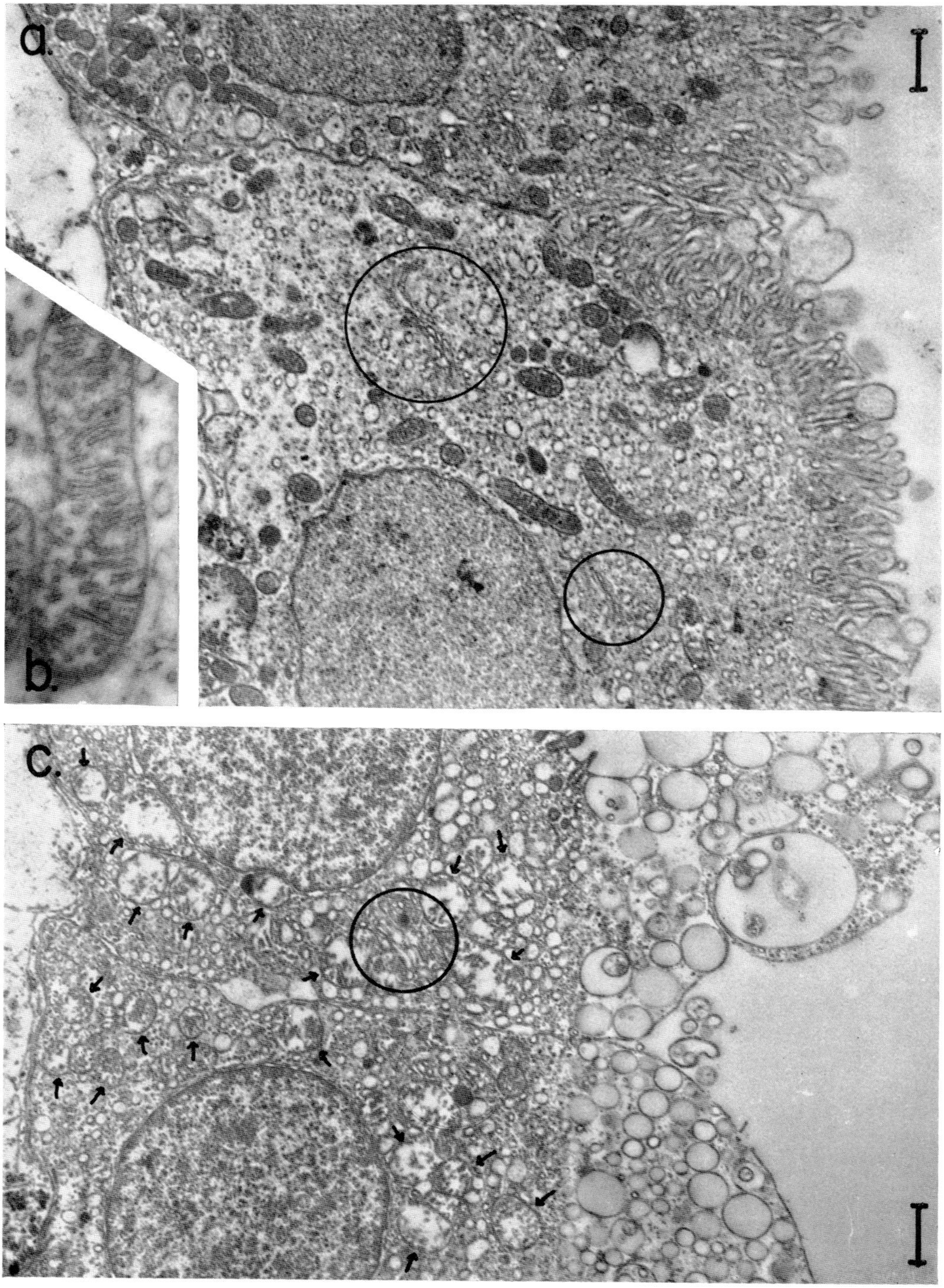
a.
b.
c.

Cells which have a good deal of endoplasmic reticulum, usually have this partly in the form of flattened cisterns which are arranged in stacks. It can be expected that the width and arrangement of the cisterns should be reasonably uniform. They may become irregularly swollen when poorly preserved.

The apparent gap between the outer and inner nuclear membranes is fundamentally a continuation of cisterns of the endoplasmic reticulum. As such, this should be of nearly uniform width.

Golgi membranes and vacuoles seem quite resistant to the vagaries of fixation, and may have good appearance even though the other organelles are severely damaged.

The plasma membrane at the surface of the cell certainly should be complete. Deep basal infoldings of this membrane, such as those found in the proximal and distal tubules of the kidney, should be separated by a nearly uniform gap.

The "ground substance" of cytoplasm should be very finely precipitated and not at all conspicuous in most cells. Irregular masses of material, coating organelles and inclusions, indicates coarse aggregation of the proteins.

The apical ends of secretory cells are apt to be more sensitive to deleterious influences than the basal ends (cf. Fig. 3). If the former seem more watery, with their contents widely dispersed by comparison, they probably have been swollen.

One also has to be aware of shrinkage artifacts comparable

FIG. 3. A comparison of reasonably well fixed (a) and (b) and poorly fixed (c) epithelium of the choroid plexus. In the well-preserved specimen this cell type exhibits a delicate "polypoid" border at the apical end. The nucleoplasm is uniformly granular. Mitochondria are small dense bodies with well-defined cristae, visible at high magnification (b). The cytoplasm is characterized by large numbers of vesicular structures, some of which belong to the endoplasmic reticular system. Groups of Golgi membranes can be found without difficulty (encircled). During poor preservation delicate apical structures of the cells undergo swelling, vesiculation, and destruction. Deeper cytoplasmic vesicles are sometimes swollen. The mitochondria (arrows) are tremendously swollen. The nucleoplasm shows evidence of coarse precipitation, particularly in a marginal aggregation of dense chromatin. Only the Golgi apparatus and some of the granules remain essentially unaltered. In general, the condition of the mitochondria and the nucleus are the most sensitive indicators of fixation damage.

to those that should be familiar to every light microscopist. With differential shrinkage, spaces can be artificially created which may be small or large. They may form between cells or around inclusions. Connective tissue is particularly apt to be torn and spaces opened.

Damage simulating poor fixation can occur in specimens when the methacrylate which is usually used for embedding undergoes polymerization (Fig. 8). Affected cells appear swollen. All cytomembranes are apt to show interruptions. Cyto- and nucleoplasm may present a flocculent appearance. "Polymerization damage," fortunately, is usually confined to the most superficial layers of tissue blocks, although it is apt to destroy freely suspended cells, or cells in tissue culture. It is discussed, and ways of avoiding it indicated, in Chapter 4.5.

Order tends to be degraded by poor preservation. Cytoplasm looks most complex when it is well preserved. Morphological beauty is a correlate of good fixation. When the good and the bad are compared there would rarely be any doubt about which was which. It is for this reason that the M.D., in particular, is again urged to begin with animal tissue to establish criteria of good preservation which he must strive for in working with human tissue.

3.7. The Osmium-Chromate Fixative of Dalton (1955)

In 1955 Dalton proposed a fixative that has theoretical interest and has enjoyed some popularity. This fixative is based on osmium tetroxide, but uses a chrome vehicle at pH 7.2 which has some buffering properties, for dichromate ions form equilibrium mixtures with chromate ions. Dalton hoped that the chromate and dichromate ions would be fixatives in their own right, and thus might be expected to preserve cell components that were not fixed by the osmium tetroxide. There was also the likelihood that chromium atoms would become attached to cell structures and thus produce the effect of an "electron stain." He thought it might be particularly effective when fixing complex lipids.

The formulation is as follows:

Stock buffer:	
4% potassium dichromate, brought to pH 7.2 with KOH	
Stock salt solution:	
NaCl	3.4%
Stock osmium tetroxide solution:	
Osmium tetroxide	2.0%
Fixative:	
Buffer	1 part
Salt	1 part
Osmium tetroxide solution	2 parts

Unfortunately the small ions of the vehicle undoubtedly penetrate tissue blocks most rapidly, and thus are apt to reach cells and their organelles in advance of the osmium complexes. Insofar as the former get there first, they probably do more harm than good so that, often, decidedly inferior fixation results from the use of this fixative. However, when cells are immediately exposed to the fixative as in tissue cultures, there is less likelihood of this happening, and excellent results may be obtained without particular difficulty.

There is probably less leaching out of cytoplasmic proteins after this fixative than after the standard mixtures. Thus, there may be a greater retention of background material, which in turn means a reduced contrast in some situations.

Although Dalton and Palade fixatives preserve the usual cytoplasmic organelles in almost exactly the same manner, there are sometimes important differences in the preservation of secretory products, which may make a comparison worthwhile.

Thorough mixing of the chrome solution, and a careful check of the pH after titration is important. Failure to do this may have disastrous consequences.

3.8. The Permanganate Fixative of Luft (1956)

In 1956 Luft demonstrated that the oxidative potential of potassium permanganate could be usefully applied to certain

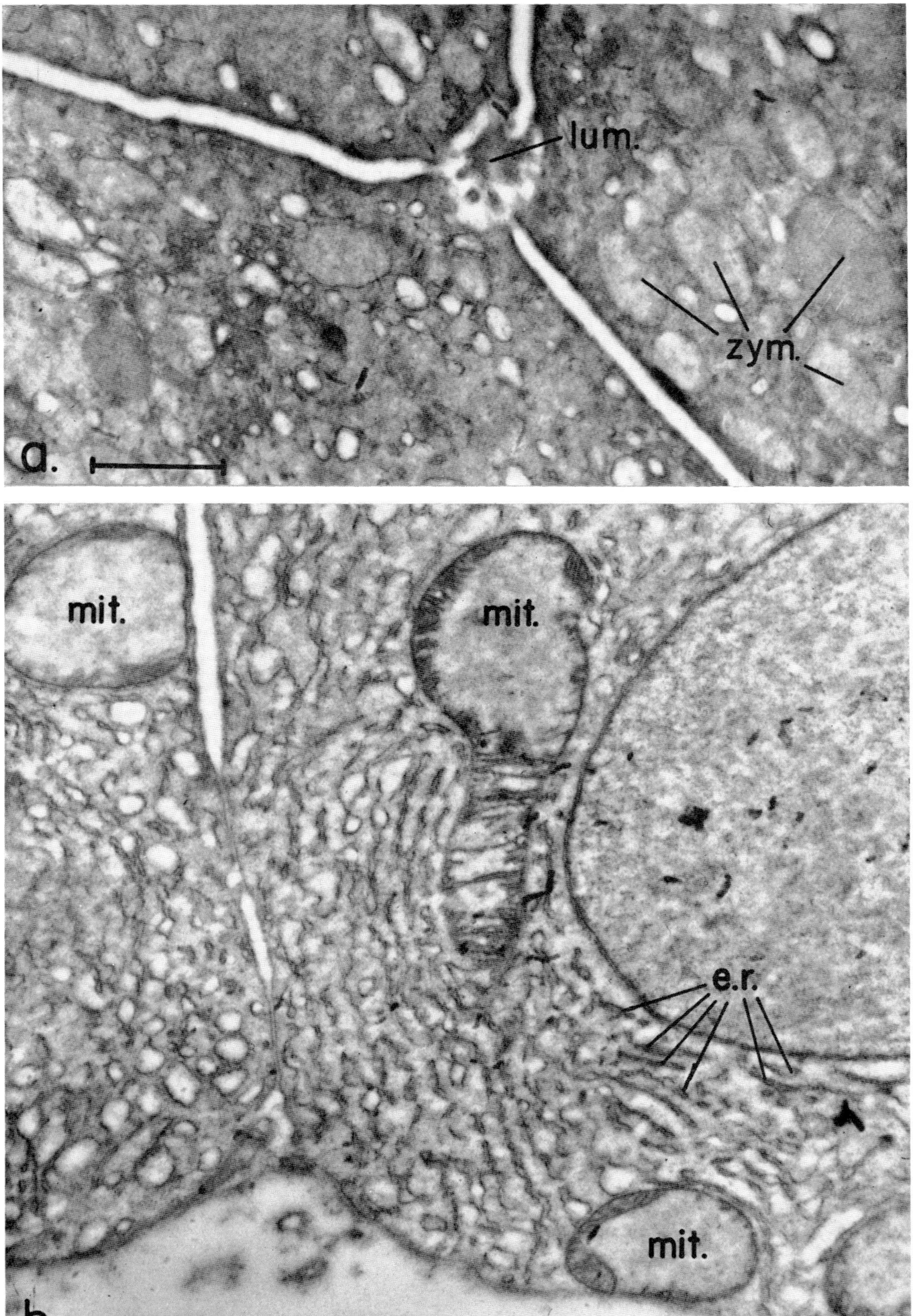
lum.
zym.
a.
mit.
mit.
e.r.
mit.
b.

problems of cytological preservation. Luft recognized, however, that the buffered mixture of potassium permanganate that he advocated was in no sense a general purpose fixative. In fact, its useful oxidative properties are largely confined to preserving lipoprotein complexes which occur in the membrane systems of a cell. Almost all other parts of a cell are destroyed. The net effect is that cell membranes appear as a tracery of thin but very dark lines against a pale background. This extreme contrast is a particularly useful property when tissue is embedded and sectioned in epoxy or polyester resins as discussed in the next chapter. To a certain extent this fixative differentiates structures within biological membranes, and has been particularly useful in studying myelin (cf. Robertson, 1957, 1958). Figure 4 indicates the general characteristics of the fixation. Whaley *et al.* (1959) have demonstrated that this fixative can be employed with advantage in studying endoplasmic reticulum in botanical material.

Luft prepared the permanganate fixative with the same buffer that Palade had used with osmium tetroxide, as follows:

Stock permanganate solution:	
Potassium permanganate (store in refrigerator in glass-stoppered bottles)	1.2%
Stock buffer solution:	
Sodium veronal (barbital)	14.7 gm
Sodium acetate	9.7 gm
Water to make	500 ml
Fixative:	
Permanganate solution	1 part
Buffer solution	1 part
Adjust to pH 7.4–7.6	

It is essential that this fixative be used at near 0° C; otherwise the oxidation proceeds altogether too vigorously, and the entire specimen literally may vanish.

FIG. 4. Cells of the rat pancreas preserved with Luft's permanganate fixative. (a) The apical region with a transverse view of the lumen (lum.), and several zymogen granules (zym.). (b) The basal region with fairly well preserved intercellular membranes, and cisternal membranes of the endoplasmic reticulum (e.r.). Mitochondria (mit.) ordinarily are greatly swollen by this fixative, as here, and the nucleus appears without differentiation.

Since this fixative is used as a progressive oxidant, time becomes an important factor. Most membrane systems are best preserved after a short time interval of from one quarter to two hours, but in special instances shorter or much longer periods can be desirable. The investigator will have to experiment to find the time interval most suited to his requirements.

In this writer's experience the fixative seems hypotonic, so that mitochondria, in particular, tend to be decidedly swollen (Fig. 4). Ways might be devised to avoid this. Tissue preserved in this way seems unduly sensitive to "polymerization damage." This can be avoided by epoxy embedding, or perhaps by using "prepolymerized" methacrylate as discussed in Chapter 4.

3.9. Formalin and Acrolein

The properties of formalin have tempted a number of electron microscopists to explore its use. Ordinarily it has only proved of value in demonstrating relationships better seen with osmium tetroxide. Indeed, tissue preserved with formalin is unduly sensitive to electron bombardment, and sometimes decomposes as it is being observed (Morgan *et al.*, 1956). Recently, however, it has become evident that tissue given a primary fixation with formalin, followed by a secondary treatment with osmium tetroxide, is reasonably well preserved for some purposes. In particular, enzymatic digestions and histochemical reactions may be carried out between the first and second fixations. Also formalin can be counted on to penetrate tissue rapidly and deeply.

This writer has been particularly impressed by some remarkable unpublished micrographs of Y. Yotsuyanagi who has been studying yeast cells which are notoriously difficult objects. Their organelles are embedded in dense, almost impenetrable, masses of RNA-granules. But these structures were beautifully revealed by using an RNA-ase digestion following fixation in 10% neutral formalin. This was carried out in the standard veronal-acetate buffer at pH 7.4. The cells subsequently were washed in buffer and exposed to the Palade fixative.

This author recently has fixed animal cells in a buffered formalin fixative, using a veronal-acetate buffer in the same final concentration as in the Palade fixative, and adding formalin to give a final concentration of 10%. Secondarily, the tissue was treated with the standard Palade fixative. Figures 5(a) and (b) demonstrate the result. Although some of the intracytoplasmic membrane systems were not well preserved so that the RNA-granules are diffusely distributed, this might be prevented by using a more sophisticated fixative than just formalin. There was also a somewhat coarse preservation of cytoplasmic proteins so that it is not possible to recommend this as a general purpose treatment.

Some botanical material which has heavy cell walls is notoriously difficult to preserve, for fixatives may not penetrate. Formalin may be effective under these circumstances. Ehrlich (1958) has used a primary fixation with neutralized formalin followed by Palade fixative with considerable success, even with microspores and pollen grains.

Formalin has also been used following fixation with osmium tetroxide in an effort to toughen tissue, and so reduce subsequent "polymerization damage" (Chapter 4.5). Its use in this connection has been mainly confined to tissue culture preparations and suspensions of free cells which are most susceptible to this sort of damage. There is, perhaps, some doubt as to whether it is truly effective.

Low and Freeman (1956) proposed a fixative combining formalin with chromic acid. Although some features of cells, such as endoplasmic reticulum and nuclear chromatin, are reasonably well preserved, other structures are lost, and the authors by no means considered that the fixative would have general applicability. Their final solution contained 3% chromic acid, 4% formaldehyde, and 0.85% sodium chloride. The pH of the mixture was about 3.2. Short fixations from ¼–½ hr proved best, after which washing seemed important. They comment that individual cells varied much in quality of preservation even though side by side.

Acrolein (acrylic aldehyde) in a number of respects resembles

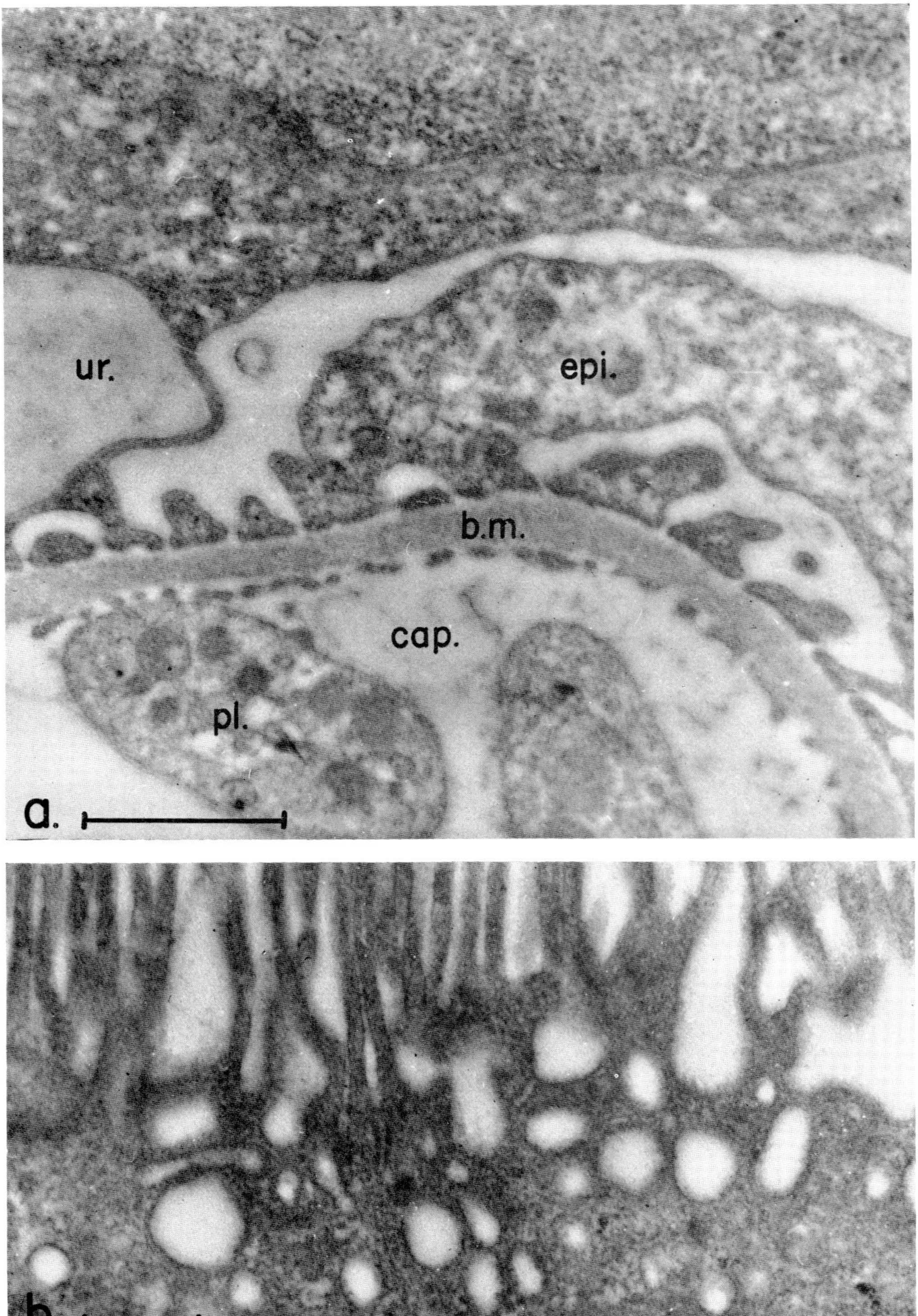
ur.
epi.
b.m.
cap.
pl.
a.
b.

formalin. Luft (1959) has demonstrated that this can be an effective fixative for conventional microscopy because it has extraordinary powers of penetration as well as being a good preservative. It may prove of considerable value to electron microscopists in some situations. For electron microscopy, fixation in acrolein must be followed by a secondary fixation in osmium tetroxide.

Luft recommends a 10% solution buffered at pH 7.4 with 0.05 *M* phosphate buffer. Tissue is fixed from 15–45 min at room temperatures. Reduced or elevated temperatures produce decidedly inferior results. Tissue is rinsed in buffer alone, and dissected or cut into blocks suitable for electron microscopy while in the buffer solution. The small blocks are washed further in the buffer, and then postfixed in buffered osmium tetroxide for about twice as long as normally would be employed. Tissue preserved with acrolein is said to be particularly sensitive to "polymerization damage" even after prolonged treatment with osmium tetroxide, and precautions, such as using prepolymerized methacrylate, should be employed (Chapter 4.5).

Solutions of acrolein are unstable and must be made up freshly every few days. Since acrolein is a violent tear gas, it is somewhat difficult to handle, and an adequate hood should be available. Although its toxicity is high, its effects are immediate, and a toxic, short-term concentration is intolerable because of the lachrymation.

Acrolein is not to be regarded as an all-purpose fixative. Although the fixation of certain organelles may approach the quality obtainable with osmium tetroxide directly, it is probably never superior to the latter. Its usefulness will be mainly to preserve complex organs which may be damaged by preparative delay or altered pressure relationships attendant to dissection,

---

FIG. 5. Rat kidney tissues fixed for 4 hr in 10% formalin buffered at pH 7.4 with veronal-acetate. This was followed by a half-hour treatment with buffered osmium tetroxide. (a) A transection through a glomerular capillary (cap.) showing parts of an epithelial cell and its feet (epi.), urinary space (ur.), basement membrane (b.m.), and platelets (pl.). (b) The apical portion of a cell of the proximal tubule showing the base of the brush border and the inclusions of adjacent areas. Although certain features of these cells are reasonably well preserved, a comparison with Figs. 2, 18, and 21 will show that some organization has been lost and, in places, relatively coarse proteinaceous precipitates have formed.

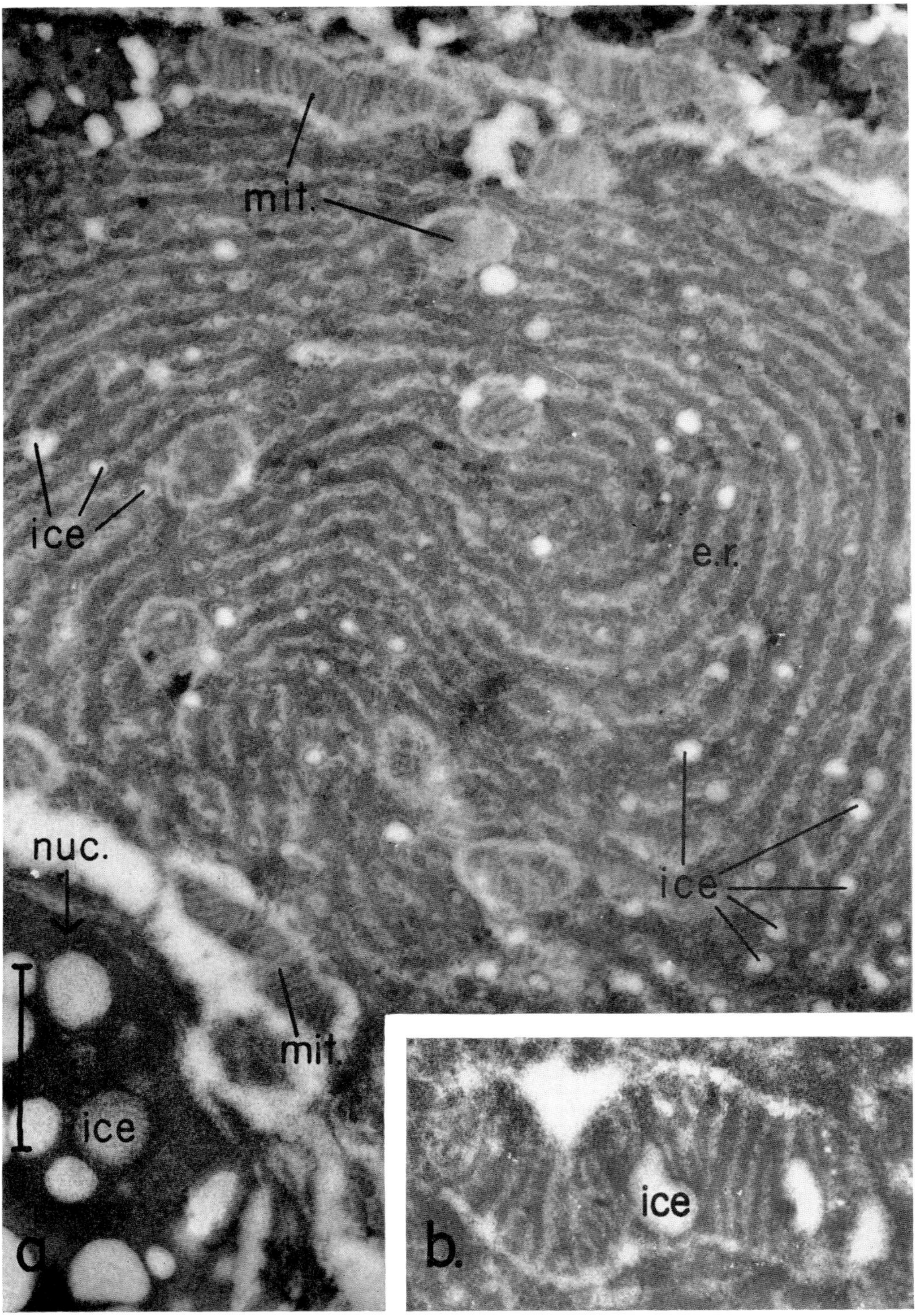
mit.
mit.
ice
e.r.
ice
nuc.
ice
mit.
a
ice
b.

but which must be dissected to isolate the structure of interest. Thus, one might fix a whole, large organ and dissect it carefully over a period of several hours, and subsequently process the tissue of interest with some confidence that the fine structure is intact. There may also be some value in the use of acrolein where it is desirable to avoid any contact with heavy metals, perhaps in future histochemistry.

3.10. Freeze-Drying

Thoughtful electron microscopists often have worried about their virtual dependence upon osmium tetroxide fixation. They have wondered what might be missed by this method, or what specific artifacts might be introduced. This, after all, is a chemical fixative, and one would like to be able to avoid chemical reactions altogether. Thus, the possibility of preserving tissue by purely physical means has been a goal sought by many, but only recently achieved in considerable measure of success by Sjöstrand and Baker (1958) who employed freeze-drying, and by Fernández-Morán (1959) using freeze-substitution (discussed in the next section). Figure 6 is of one of the formers' preparations. Note that specimens which have been frozen-dried and subsequently embedded in methacrylate have had their lipids removed by the solvent action of the methacrylate monomere. The membrane systems of cells, thus, are ghostly in appearance, visible mainly insofar as they are outlined by other structures.

Sjöstrand and Baker discovered that the real difficulty in preparing frozen-dried material for electron microscopy comes when the dried tissue is to be infiltrated with plastic monomere. The monomere does not wet all parts of a dried cell if the latter contains air. Therefore frozen-dried tissue cannot simply be dropped

---

FIG. 6. (a) Frozen-dried mouse pancreas prepared by Sjöstrand and Baker (1958). A phosphotungstic acid "stain" was used to improve contrast. Note that lipids have been extracted so that the membranes of the nucleus (nuc.), the mitochondria (mit.), and the cisterns of the endoplasmic reticulum (e.r.) appear as pale lines, while proteinaceous material survives the treatment, and subsequently is densely stained. Ice artifacts (ice) are particularly prevalent in the nucleus, but are not serious in the cytoplasm. (b) An enlarged view of a mitochondrion.

into methacrylate. The gaseous films which remain cause great physical damage to the tissue during subsequent polymerization of the embedding medium. These difficulties can be avoided if infiltration is completed in a high vacuum. Since butylmethacrylate has a considerable vapor pressure when it is in liquid form, it is not possible to obtain a good vacuum in its presence unless it is first frozen and an equilibrium mixture with its own melt used. This can be achieved at a temperature of approximately $-120°$ C.

Previously frozen-dried tissue is placed upon a chunk of frozen methacrylate monomere within a vacuum chamber. The system is exhausted, maintaining the low temperature until a high vacuum is reached. Then the system is allowed to warm up slowly so that the methacrylate monomere is in equilibrium with its own solid and the vapor pressure remains low. The tissue sinks into the melt and is completely wet by the monomere so that excellent infiltration occurs. After this, the vacuum can be released by admitting dry air, and the system warmed to room temperature. (It may be suggested that in the future this rather cumbersome procedure might be eliminated by finding some organic liquid which would wet frozen-dried tissue on the one hand, and be miscible with methacrylate on the other. It then might be used as an intermediate step at room temperature.) Subsequent polymerization is carried through in the usual way as described in the next chapter.

If material is infiltrated in the manner described, it appears that commonly accepted methods of freezing and drying are adequate to give well preserved tissue. The tissue to be frozen should, of course, be a very minute piece so that it can be frozen with extreme speed, thus avoiding the formation of ice crystals. The few small ice pockets that may appear (Fig. 6) do not seriously detract from the results. The tissue can be frozen by plunging it into a bath of isopentane cooled with liquid air or liquid nitrogen. Tissue should not be plunged directly into liquid air or nitrogen for gaseous films then form around it. These gaseous films function as insulating layers which prevent the rapid transfer of heat from the specimen to the bath.

It is likely that dry ice temperatures (—72° C) also will prove adequate for freezing small bits of tissue. A container of ethyl ether into which the specimen is plunged may be precooled with a dry ice and alcohol mixture. (Absolute alcohol can be used as the specimen bath, but tends to absorb water so readily that the formation of an annoying sludge of ice crystals is almost inevitable.) After freezing, tissue is transferred to a precooled container which is subsequently evacuated.

The design features of suitable vacuum systems are well known and have been discussed by many authors (cf. Bell, 1956 and Harris, 1954). They include the use of a good modern diffusion pump and a "cold finger" which can be filled with liquid nitrogen. The latter is introduced close to the specimen to be dried. The specimen is maintained at a temperature higher than that of the cold finger, and water vapor distills from the one to the other in the same manner, fundamentally, as water leaving food in a freezer collects on its coils. It is generally agreed that successful drying must be completed at temperatures below —30° C. There are many who think that the drying should be at much lower temperatures than this. Water still has an appreciable vapor pressure at —72° C, the temperature of a dry ice-alcohol bath, so that this is a convenient temperature to work at.

In designing vacuum systems one other point deserves mention. Physicists emphasize that all connections must be as short and as wide as is practically possible. One cannot use small bore tubing to fasten the components together without making a tremendous sacrifice in the efficiency of the apparatus.

Vacuum gauges are one of the major expenses in vacuum systems. While there can be no question but that good gauges are desirable, if one has a relatively simple glass system beyond the diffusion pump, this can be expected to work reliably, and one might empirically determine the time necessary to dehydrate specimens. One also can use the discharge of a relatively cheap, high frequency induction coil (Tesla coil) as an approximate indication of high vacuum. A good vacum exists when the glow discharge nearly or totally disappears.

Since materials which are preserved by freeze-drying presumably retain their original solubilities, special difficulties arise during sectioning. As will become apparent in Chapter 5, sections normally are floated upon aqueous solutions, which undoubtedly would leach water soluble materials out of frozen-dried sections. Perhaps because of this, Sjöstrand and Baker (1958), reported that RNA-granules had disappeared from their preparations. In the future it will be necessary for people to show ingenuity in collecting specimens without wetting them with water. One might return to the techniques of the earliest days of ultrathin sectioning when dry sections were picked up with hairs mounted in handles, and tacked down on screens. One can suppose that more efficient ways will be devised.

Successful frozen-dried sections are not the immediate cure-all for all problems. Indeed, material prepared in this way has such low intrinsic density that whole new techniques of "electron staining" will have to be developed before they will be generally useful. It is quite obvious from the work of Sjöstrand and Baker that "unstained" frozen-dried sections must be relatively thick in order to have sufficient contrast to make a picture that is at all satisfactory. The thickness of the sections precludes high resolution micrographs. At the present time, therefore, an unstained frozen-dried section is useful mainly for demonstrating the reality of cytoplasmic features that we can study much more effectively after osmium tetroxide fixation. For the future, of course, the way is now prepared for the development of a histochemistry at electron microscopic levels, and the successful frozen-dried sections of Sjöstrand and Baker mark a technical milestone.

3.11. Freeze-Substitution

Freeze-substitution differs from freeze-drying in several important respects. By means of suitably chosen intermediates a piece of tissue can be dehydrated and infiltrated with an embedding medium at low temperature without there ever being a gas phase present. Instead of requiring a highly efficient pumping system, one needs only suitable cold baths and boxes. It offers the possible advantages of exposing tissue to fixation and staining agents during the infiltration process. It may even prove possible to incorporate photosensitive material within tissue blocks and so make feasible the marriage of autoradiography and electron microscopy. It promises to be more useful than freeze-drying for histochemistry since it has much greater flexibility.

There would be little point is discussing freeze-substitution in relation to electron microscopy except for the very recent work of Fernández-Morán (1959) which is still in an exploratory phase of development. This investigator supplied the present writer with many details of his techniques, which are incorporated in what follows. At the present time it is difficult to decide what will be of permanent value, and the reader should regard this section as no more than a general guide, suggestive of what can or may be done. If the reader intends seriously to invade this field he will want up-to-the-minute information from Fernández-Morán, and will have to show much originality in working out appropriate practices for himself.

Fernández-Morán eventually hopes to work with temperatures below —130° C where one can be fairly certain that he has only vitreous water in his specimen providing that it was frozen sufficiently rapidly. He expects that it will prove possible to infiltrate with a plastic monomere and catalyze polymerization at such a low temperature, but for the present there are still severe technical difficulties. However, it is quite feasible at the present time to carry out all the necessary operations at —72° C (the temperature easily reached by dry ice mixtures) or —90° C (achieved fairly easily with commercial refrigerator chests). If suitable precautions are taken to avoid "ice-crystal seeding" in all of the solutions used, and if the tissue is not excessively hydrated, it is possible, at least in some instances, to infiltrate successfully at these higher temperatures without having the tissue seriously affected by ice crystals. Thus, freeze-substitution becomes a practical and potentially very valuable technique.

Fernández-Morán has been exploring a variety of methods for freezing tissue. On the one hand he has been using helium II. This substance is notorious for its well nigh incredible mobility and for the speed with which it transfers heat. Both of these properties may prove invaluable in special instances, but certainly are not generally necessary for tissue work. For much work conventional freezing in isopentane or, better, Freon 22 chilled with liquid nitrogen will suffice. More interesting is the possibility of using heavily glycerinated tissues which may not require rapid freezing at all. Thus, living chick embryos can be infiltrated with about 20% glycerine absorbed through their external membranes while the temperature is slowly dropped to about 0° C. At about this point the heart stops, then tissue is removed from the embryo and is passed through a series of precooled glycerol solutions until it is equilibrated with 60% glycerine at −25° C. The tissue finally can be frozen safely at lower temperatures without ice crystal formation. To some extent at least, this procedure is applicable even with adult mammalian tissues, for cold glycerine solutions can be perfused via circulatory pathways while either the whole animal or the part to be studied is chilled by external packing.

Methyl cellosolve appears to be the substance of choice which first will substitute for the tissue water and glycerine, and then can be replaced with methacrylate (and related monomeres) in the temperature range of −72° to −90° C. For special purposes other intermediates such as acetone or alcohols can be used. When working at −90° C, and particularly at even lower temperatures, 10–50% ethyl chloride may be added to lower the eutectic point and maintain adequate fluidity.

Fernández-Morán emphasizes most importantly, that ice-crystal "seeding" must be avoided at low temperatures, and extraordinary precautions must be taken to avoid this with all solutions that are used. Mainly the problem is to avoid water crystals that result from trace quantities of water in even the best of commercially available, so-called anhydrous reagents. It is necessary to remove traces of water by exposing all reagents to Linde 4A molecular sieves. Then particulate matter is removed by filtering the anhydrous compound with a sintered bacteriological ultrafilter. In some instances it is also necessary to have vacuum-redistilled the reagent before this. Reagents prepared in this way subsequently are kept under vacuum to avoid gas absorption. Before chilling, the reagents to be used are put in sealed chambers of a design such that they can be connected to the specimen chamber at will, and transfers made without breaking the seal. Thus, water vapor is never allowed to condense upon and contaminate the reagents or specimen.

The design of an apparatus which permits fluid transfer in a specimen chamber without water vapor contamination is not rigidly determined. It is convenient to use separatory funnels equipped with Teflon stopcocks as containers for the bulk fluids. They may be connected to the specimen chamber with Teflon tubing. Alternatively they may be fitted with nipples of a size to fit standard hypodermic needles. If the specimen chamber has Teflon sidewalls, entrance and exit ports can be made by sticking wide-bore hypodermic needles through these. The needles are themselves equipped with stopcocks. Thus, different reagent bottles can be connected in turn, breaking no more than the one tiny seal for a brief interval of time.

Fernández-Morán suggests that a simple "model system" be incorporated in the specimen chamber which can serve to give a clue as to when substitution is complete. Basically this is no more than a short length of thin-walled capillary, a little bit longer than the specimen is thick. This is filled with a colored salt solution which is frozen with the specimen. The retreat of the interface can be followed during substitution, and serves to indicate its rate. If riboflavine phosphate is the dye used, dissolved in a glycerinated salt solution resembling the specimen, one can tell whether or not crystalline ice exists, for in the vitreous state this substance shows a brilliant yellow ultraviolet fluorescence, while crystals have an orange phosphorescence.

Substances such as uranyl acetate, lanthanum nitrate, and platinic chloride can be dissolved in methyl cellosolve, and thus "staining," and a degree of fixation, can be accomplished during the substitution steps. If osmium tetroxide is to be used, apparently it is better to dissolve this in cold acetone, and infiltrate first with this before substituting methyl cellosolve. Osmium tetroxide in acetone is reasonably stable if kept at dry-ice temperatures.

It is at these low temperatures, too, that silver halides can be incorporated into tissue blocks which contain radioactive tracer material. A unique degree of molecular stabilization is achieved, and the tracer itself is immobilized in a position equivalent to that which it had in the living organism. Moreover, elimination of interfering local chemical activity enhances the possibility of very precise localization of the emission sites. Fogging is inappreciable, even during an exposure period of weeks, if the temperature is kept below $-30°$ C. Eventually the *latent image* grains can be "gold plated." The details of these procedures, however, are not yet ready for release as this is being written, and this is more a hope for the future than a presently practical method.

After the tissue has been completely infiltrated with methyl cellosolve it is then ready for exchange with plastic monomeres. Fernández-Morán recommends that appreciable quantities of acrylates be added to the usual methacrylate mixture. Thus, his recommended formulation is butylmethacrylate, 80 parts; methylmethacrylate, 10 to 15 parts; methylacrylate, 5 to 10 parts; plus, as catalysts, 1% benzoyl peroxide and 0.2% benzoin. If one wishes to infiltrate in the $-90°$ C range instead of at $-72°$ C the concentration of methylacrylate should be increased to perhaps 20%. Infiltration is carried on over a 2–14-day period using four changes of the monomere mixture.

It will ordinarily seem safe to raise the temperature of the specimen to perhaps $-30°$ C during this infiltration process. If this is done, ultraviolet polymerization of the plastic can be accomplished very easily with a simple fluorescent ultraviolet lamp, and one need not be seriously concerned about thermal effects and infrared radiation. If it is necessary to polymerize at $-72°$C, or at lower temperatures, more elaborate ultraviolet equipment will prove necessary. Then, to avoid heating effects, the lamp will have to be mounted outside of the cold box, and the radiation must be filtered to eliminate infrared light.

Material which has been fixed at one stage or another now can be sectioned in a conventional manner. Material which is unfixed, and which perhaps may be wanted for histochemical investigation, can be sectioned if necessary in a dry, cold-box, and the sections collected on an appropriate Freon. If there is no subsequent fixation, sections of this sort are apt to decompose badly when bombarded by electrons in the microscope unless a cold stage (such as is available for the Siemens instrument) is used during microscopy.

It is obvious that this general method of freeze-substitution offers a flexible technique adaptable to a variety of different research problems. Much can be expected of it in years to come. It is not a routine method as yet, however, and the investigator who would begin work in this field will have to show resourcefulness in working out technical details. If the investigator is content to work at —30° C he can buy a freezer without great expense, and employ relatively simple transfer equipment. If he would work at —90° C, a costly cold-box (available from the Harris Refrigeration Company, Boston) is the principal item of equipment to be purchased. Transfer apparatus will have to be custom made, but need not require very complicated glass blowing or machine work. Freeze-substitution below this temperature obviously is something for the future. Not only does the equipment become a problem, but different solvents and new embedding plastics must be sought. There is some indication that suitable embedding plastics can be found, and that appropriate catalysts will allow polymerization even at extremely reduced temperatures.

Literature Cited

Bell, L. G. E. (1956). *In* "Physical Techniques in Biological Research" (G. Oster and A. W. Pollister, eds.), Vol. III, Chapter 1. Academic Press, New York.

Caulfield, J. B. (1957). *J. Biophys. Biochem. Cytol.* **3,** 827.

Claude, A., and Fullam, E. F. (1946). *J. Exptl. Med.* **83,** 491.

Dalton, A. J. (1955). *Anat. Record* **121,** 281.

Dalton, A. J., Kahler, H., Striebich, M. J., and Lloyd, B. (1950). *J. Natl. Cancer Inst.* **11,** 439.

Ehrlich, H. G. (1958). *Exptl. Cell Research* **15,** 463.

Fernández-Morán, H. (1959). *Science* **129,** 1285; and Proc. Elec. Micro. Soc. Amer., *J. Appl. Phys.* **30.**

Harris, R. J. C., ed. (1954). "Biological Applications of Freezing and Drying." Academic Press, New York.

Low, F. N., and Freeman, J. A. (1956). *J. Biophys. Biochem. Cytol.* **2,** 629.

Luft, J. H. (1956). *J. Biophys. Biochem. Cytol.* **2,** 799.

Luft, J. H. (1959). *Anat. Record* **133,** 305.

Morgan, C., Moore, D. H., and Rose, H. M. (1956). *J. Biophys. Biochem. Cytol.* **2** (Suppl.), p. 21.

Palade, G. E. (1952). *J. Exptl. Med.* **95,** 285.

Palade, G. E. (1956). *Proc. Third Intern. Conf. Electron Microscopy, London, 1954,* p. 129, Roy. Micro. Soc., London.

Palay, S. L., and Palade, G. E. (1955). *J. Biophys. Biochem. Cytol.* **1,** 69.

Pease, D. C., and Baker, R. F. (1948). *Proc. Soc. Exptl. Biol. Med.* **67,** 470.

Pease, D. C., and Baker, R. F. (1950). *Am. J. Anat.* **87,** 349.

Porter, K. R. (1950). *Anat. Record* **106,** 145.

Porter, K. R. (1953). *J. Exptl. Med.* **97,** 727.

Rhodin, J. (1954). Correlation of Ultrastructural Organization and Function in Normal and Experimentally Changed Proximal Tubule Cells of the Mouse Kidney. Anat. Dept., Karolinska Institutet, Stockholm.

Robertson, J. D. (1957). *J. Biophys. Biochem. Cytol.* **3,** 1043.

Robertson, J. D. (1958). *J. Biophys. Biochem. Cytol.* **4,** 39 and 349.

Sjöstrand, F. S., and Baker, R. F. (1958). *J. Ultrastructure Research* **1,** 239.

Whaley, W. G., Mollenhauer, H. H., and Kephart, J. E. (1959). *J. Biophys. Biochem. Cytol.* **5,** 501.

Zetterqvist, H. (1956). The Ultrastructural Organization of the Columnar Absorbing Cells of the Mouse Jejunum. Dept. Anat., Karolinska Institutet, Stockholm.

4. Embedding

4.1. Introductory Remarks

The earliest attempts at ultrathin sectioning failed principally because of an inadequate embedding medium. Pease and Baker (1948) achieved a measure of success because they combined the tough properties of the plastic "Parlodion" with the hardness that could be obtained with a carnauba-paraffin wax mixture. Material "double-embedded" in this way sections well. It seemed necessary at the time to extract the wax component with an organic solvent after sectioning in order to gain the contrast between specimen and background necessary for satisfactory micrographs. Unfortunately this reduced the support of the specimen to such an extent that surface tension forces came into play and destroyed much of the fine structure present. It has since come to be recognized that one must never chemically extract the embedding medium.

Newman *et al.* (1949) introduced butylmethacrylate embedding which since has become the standard method. This was a fortunate choice of material quite apart from its cutting characteristics, for methacrylate plastics partially decompose in an electron beam. Some of the decomposition products volatilize, which removes "background" material, and greatly enhances specimen contrast (Fig. 8). This is a very important property from the point of view of practical electron microscopy, yet it is now realized that one must pay for this gain in contrast by accepting an artifact that occurs at macromolecular levels. Apparently once again surface tension forces are the villains. As methacrylate is heated by the electron beam and decomposes, liquid interfaces are formed transiently, and fine order detail in the specimen may collapse. None the less, convenience and adequacy for most purposes, continue to make this by far the most useful and popular embedding medium.

It was work with epoxy resins, begun by Maaløe and Birch-

Andersen (1956), that disclosed the limitations of methacrylate embedding, for these resins do not liquefy or decompose in the electron beam. Surface tension forces are not active, and macromolecular structure is preserved intact. In this situation the price one pays is a very low specimen contrast, so low, indeed, that usually electron "stains" must be used to visualize anything. This is too expensive a price to allow this method to be regarded as a general purpose technique, although it has proved very important in special investigations.

Kellenberger and co-workers (1956) introduced polyester resins. Material, as currently embedded, gains some contrast as it is bombarded, yet there is little or no loss of specimen support so that very fine detail is preserved. This embedding material so far has not been given the rigorous tests with highly organized macromolecular systems (such as muscle fibrils or nerve myelin) that have demonstrated the qualities of epoxy resins. If it preserves as well as the latter resins, the improved contrast may make it the embedding medium of choice for high resolution electron microscopy. So far it is not possible to handle it as easily as methacrylate, and the contrast is decidedly inferior. Thus, it is not yet the ideal, all-purpose embedding medium.

4.2. Dehydration

Methacrylate monomeres are miscible in ethyl alcohol so that most commonly a graded series of this alcohol is used for dehydration purposes. As indicated in the previous chapter, though, electron microscopists have learned to be concerned about materials being leached out of osmium-fixed specimens, even while they are in the fixative, and also while being dehydrated. It is, therefore, commonly accepted that dehydration should be completed as rapidly as possible. In some laboratories also it is the custom to dehydrate at near zero temperatures. Naturally this would have little meaning unless the fixation also had been completed in the cold.

When tissue blocks are first removed from the fixative they probably should be given a quick rinse with water, for excessive quantities of osmium tetroxide react with alcohol. In this writer's laboratory de-ionized water has been used for some time on the assumption that most proteins would be less soluble in this than in salt solutions. However, other laboratories do advocate salt solutions or the standard veronal buffer solution of the fixative prepared without the addition of osmium tetroxide. Little more than a surface washing of the specimen and a flushing of the container are necessary before adding dilute alcohol.

Useful tissue blocks are so tiny that the exchange of water and alcohol is very rapid. In this laboratory we pass blocks of the usual size through 50%, 75%, and 95% alcohol, allowing about 3–5 min in each bath. Thus, the specimens ordinarily reach absolute alcohol within 15 min. As an experiment in this laboratory we have, on occasion, gone directly from fixative to absolute alcohol without apparently harming the tissue in any way. If we had the courage of our convictions we perhaps would do this routinely.

It is sometimes desirable to dissect or trim a slab of tissue during dehydration. This can best be done in the 75% or 95% bath. Thus, advential layers can be removed to make subsequent sectioning easier, or a thin sheet of material that was fixed *in situ* can be cut up into small blocks.

Tissue must be thoroughly dehydrated before passing it to methacrylate monomere, so several changes of absolute alcohol are suggested. Care should be taken that the container, and particularly its cork or lid, do not retain traces of water which eventually will contaminate the methacrylate. There is little point in hurrying this process of final dehydration for the tissue will be exposed for hours to unpolymerized methacrylate which will act as an excellent organic solvent that cannot be avoided. In our own laboratory we usually use two short baths of absolute alcohol (about 5–10 min each) followed by a longer one (15–20 min). Finally, methacrylate monomere is substituted, and the tissue blocks soaked in this for 10 min or longer before transferring the tissue to individual gelatin capsules containing fresh monomere (with catalyst) for final polymerization. If tissues are to be stored before blocking, this is done in methacrylate monomere.

From time to time agents other than ethyl alcohol have been proposed for dehydration. Acetone has been recommended by several investigators, yet in this writer's experience severe specimen damage from incomplete dehydration is a danger, and it is suggested that it be used cautiously if at all. The difficulty lies in getting and keeping anhydrous acetone. This writer also has some reason to distrust dioxane. In addition he has tried "Cellosolve" and dimethyl formamide. These latter substances did not seem to damage specimens, but offered no advantages over ethyl alcohol. Both dioxane and dimethyl formamide are somewhat dangerous to handle.

It has been suggested that methyl alcohol be used as a dehydrating agent, apparently because of its lesser solvent action. Although this writer has not had experience with it, it seems unlikely that there is an important advantage.

The small size of tissue specimens makes it impractical to move them from one bath to another. Instead, fluids should be poured out of the container and replaced with others. Of course, ordinary vials can be used. Much better, however, are the short, wide-mouth weighing bottles (Parr type, 4 ml, 23 × 24 mm) that have outside ground-glass caps (illustrated in Fig. 1). Specimens ordinarily are fixed in one of these, and stay in the original container until ready for final transfer to individual gelatin capsules just before polymerization.

It is convenient to use gelatin capsules as individual containers for polymerization. Since size "00" has the proper diameter to fit directly into the chuck of a Porter–Blum microtome, it is the size most commonly used in this country. The capsules are loaded for methacrylate embedding with the mixture of monomeres of one's choice, the catalyst, and the specimen which previously has been infiltrated with monomere. It is then polymerized with heat or ultraviolet radiation.

Loaded gelatin capsules are most simply labeled by writing on the surface with ink. A better label, written on paper, can be included within the capsule and so become embedded in the plastic. Labels of this sort must be of such a small size that they can be fitted to the curvature of the side walls of the capsule, and they should be placed within the capsule before its other contents.

If a specimen does not require particular orientation, it is most convenient to mount the entire hardened contents of a capsule directly in the microtome chuck, and only trim the head region to expose the tissue. Thus, the plug of methacrylate serves as the support arm carrying the specimen. This can be stiffened if a short length of heavy-walled glass tubing is embedded in the plastic. The tubing should be of a diameter to fit loosely in the capsule, but not to protrude into the curved tip of the capsule where the specimen will lie. Although some laboratories make a practice of reinforcing their capsules in this way, it certainly is not generally necessary.

4.3. Desirable Characteristics of Methacrylate Blocks

It is clear that in the foreseeable future most ultrathin sectioning will continue to be done with methacrylate embedded material. In the decade of its use for this purpose the problems of handling it have become fairly well understood. These will be discussed at some length in the next section of this chapter. Of course, the end result that one seeks is a block with excellent cutting characteristics. Hardness is one of the physical properties that must be considered. Perhaps even more important are factors of elasticity, plasticity, and homogeneity.

Block hardness is easiest to deal with. Certainly much of the early difficulty with methacrylate embedding was because people made blocks that were altogether too hard. To cut properly, the blocks should be soft enough to be indented with a finger nail. When squeezed hard, one should have the definite impression of compression. When they are being trimmed with a sharp razor while being observed with a low power binocular dissecting microscope, it should be possible to cut thin shavings with almost no feeling of resistance. It is surprising how soft blocks can be and yet cut very successfully. Do not worry too much about supporting a flexible block of plastic, for this can be taken care of in other ways than by using a harder plastic. The beginner is certainly apt to err in trying to cut too hard a block rather than too soft a one.

One can have several blocks that would all be described as equally hard, and yet there can be considerable differences in the way they cut. Now we are dealing with physical properties that are more subtle than just the degree of hardness. The most difficult type of block to section (sometimes really impossible) seems not to be truly homogeneous throughout. If one attempts to make shavings with a sharp razor these will have a frosted appearance rather than being glassy. It is as though the methacrylate was gritty or granular. The plastic lacks the rubbery quality that is highly desirable. This author has experienced this poor embedding only when heat polymerization has been used. Sometimes whole batches of capsules have been affected in this way, and one might suppose a contaminant was the cause, or perhaps, an incorrect mixture of plastic and catalyst, or improperly controlled temperature. However, the difficulty may involve only some capsules in a group, and then its cause is obscure. Fortunately one can go for long periods without being plagued by this effect, and it is not a problem when ultraviolet polymerization is employed.

A desirable block is one from which glassy and rubbery shavings can be cut with a sharp razor. The shavings should be homogeneous in character; they should have coherence and strength. If a shaving is pulled, it should stretch considerably before breaking. It is only moderately elastic in a true sense for there is little quick return after stretching. Thus, it has plasticity. With experience one can tell pretty well during the trimming whether or not the block will section easily. There will be very much trouble with "chattering" (Fig. 17) in sectioning a poor block so that only very small sections at best can be cut, while from a good block, sections four times as large will cut smoothly without any chatter at all.

4.4. Methacrylate Embedding

The hardness of the final blocks depends mainly upon the particular methacrylate or mixture of methacrylates that is used.

Polymerized butylmethacrylate is quite soft, and industry uses sheets of this material for such flexible things as belts and handbags. It is hard enough, however, to use in pure form to produce blocks which section satisfactorily. Ethyl- and methylmethacrylate, when polymerized, produce hard substances, the Lucite and Plexiglas products of industry. These are altogether too hard for sectioning purposes, but all three monomeres are compatible, one with another, and mixtures give plastics of varying and controllable hardness. Most investigators find pure butylmethacrylate blocks to be undesirably soft. They therefore add anywhere from 10 to 30% of one of the smaller monomeres to the butylmethacrylate before polymerizing. In this writer's laboratory we routinely use 10–20% ethylmethacrylate (with ultraviolet polymerization). Methylmethacrylate has been most commonly used, presumably because of its cheapness and availability, since this is the principal plastic of commerce. But it seems theoretically desirable to use ethylmethacrylate instead. In the quantities used its greater cost is without significance.

Several different catalysts are used commercially for polymerizing methacrylates. The biologist wants one that is effective at relatively low temperatures. For this reason a mixture known by the trade name of "Luperco CDB" (the catalyst, 2,4-dichlorobenzoyl peroxide, with a plasticizer, dibutyl phthalate) achieved much popularity originally for it was active at the lowest possible temperature (40° C). This mixture has to be kept refrigerated, however, and moisture condenses on it as it is being used, a most undesirable feature. Also it is in the form of a paste which makes weighing difficult. Nowadays it seems altogether better to use pure benzoyl peroxide (or 1,2-dichlorobenzoyl peroxide) as the catalyst of choice. This material is available in dry granular form, and it is not deliquescent. It can be stored for many months at room temperature. Eventual decomposition does occur so that it ceases to be properly effective. Therefore, small quantities should be purchased with the expectation of replacing the catalyst once or twice a year. The main supply should be kept tightly stoppered in storage, or even in a desiccator, to increase its life. A small container of less than an ounce will serve daily use, and its contents should be discarded and replaced when there begins to be doubt about its efficacy.

Polymerization can be induced either by mild heat or by relatively long wavelength ultraviolet radiation. This writer much prefers the latter method for reasons that will become apparent, but heat is the more traditional technique.

When heat polymerization is employed, 2% benzoyl peroxide (or Luperco CDB) should be fairly accurately weighed and added to the bulk methacrylate. It dissolves somewhat slowly, and this must be completed before the solution is added to the embedding capsules. Overnight heating at 50° C will start polymerization. Since it may not complete it, it is better practice to allow two days for this.

Methacrylate monomere is supplied by the manufacturer with a hydroquinone "inhibitor" in it. It originally was assumed that this had to be removed before attempting polymerization. *This is quite unnecessary,* however*. The presence of the inhibitor does not affect the final properties of the block, but only influences the rate of polymerization.

Almost from the beginning of methacrylate embedding there have been occasional comments that polymerization was achieved by ultraviolet radiation. The nature of the source has rarely been stated, however, and this writer knows of many investigators who have been discouraged by failure. Ironically, it seems to be the cheapest sort of fluorescent lamp that is most effective, while a high quality arc or mercury vapor lamp often will *not* induce

* Removing the inhibitor involves shaking the monomere with a small quantity of 30–50% sodium hydroxide solution in a separatory funnel. After two or three treatments the solution should be colorless, and this is followed by a number of washings with distilled water to remove the alkali. So long as alkali remains, the drain water will produce a feeling of slipperiness when rubbed between the fingers. It is then necessary to remove the last water completely. If the methacrylate is filtered through several layers of paper, all but traces of the water will be absorbed by the paper. Following this, a drying agent such as anhydrous calcium chloride or sulfate can be used. Finally, any particles of this must be removed by filtration before the monomere is used for embedding. If the water is not thoroughly removed, serious bubbles may appear during polymerization. Thus, removing the inhibitor is not only a laborious process, but also a potential source of trouble.

polymerization. These apparently produce radiation of too short a wavelength. Polymerization seemingly is carried on most efficiently by quite long wavelength ultraviolet light.

For about three years this writer has been using Westinghouse Fluorescent Sun Lamp Bulbs, #5TT12 with completely satisfactory and consistent results. This is a low cost, 24-in. tube which fits into standard fluorescent fixtures. The bulbs have a low current consumption, do not liberate much heat, and have a long life. They can be run continuously for many days. Some other types of fluorescent "black light" bulbs will work equally well.

Loaded capsules to be polymerized are simply lined up in a rack about 1 in. away from the light source. Naturally, the rack must be designed so that most of each capsule (particularly its base) receives direct radiation. One can expect much overnight polymerization even though the "inhibitor" is left in the methacrylate. But two days should be allowed for complete polymerization.

The amount of catalyst is not particularly critical when ultraviolet polymerization is employed. In this writer's laboratory, it is our practice simply to add a few grains (8–12) of benzoyl peroxide to each loaded capsule before sealing it. These grains are picked up on a forcep's tip previously wet by dipping in methacrylate. It is necessary to mix the contents of the capsules after the catalyst has dissolved or layering may result in uneven polymerization.

There are several reasons why this writer now prefers ultraviolet to heat polymerization. It is his experience that it produces a substantially superior block. Clearly, the physical properties of the final plastic are somewhat different after the two treatments. After exposure to ultraviolet radiation the plastic is more "rubbery." This is a desirable feature. The blocks are softer, but this can be compensated for by slightly increasing the proportion of ethylmethacrylate. Blocks are also more uniform. (With heat there is apt to be considerable variation for unknown reasons, and one sometimes gets a nonuniform, almost granular polymerization that is almost impossible to section. We have not had this happen when using ultraviolet radiation.) Trapped bubbles not infrequently mar heat-polymerized blocks, but ordinarily are not a serious problem when ultraviolet polymerization is used.

If one does experience serious numbers of bubbles trapped around the specimens in polymerized blocks he should consider the possibility of water contamination. However, there are certain tissues, particularly dense connective tissues, which are prone to produce bubbles in spite of any amount of dehydration. It has been said that these can be reduced or eliminated by partially degassing the methacrylate monomere before use. This can be done with an aspirator pump, and, in any case, the vacuum should not exceed 10 mm Hg, or the methacrylate monomere will start boiling. Even when large bubbles are present, however, these do not ordinarily affect the embedded tissue. If the cavities weaken the support of the block, they can be filled before sectioning by a short soaking of the trimmed capsule in hot wax as discussed in Section 4.10 of this chapter.

Moore and Grimley (1957) believe that if dissolved oxygen is replaced by nitrogen in the monomere, not only will bubbling be reduced, but the blocks will be more uniform after heat polymerization. They, therefore, advocate storing degassed methacrylate under a dry nitrogen atmosphere. This prepared monomere is then used for the preliminary soaking of the specimens, and for loading the capsules. Small holes are punctured in the tops of each capsule, and the capsules are placed upright in small vials in a vacuum desiccator. The capsules then are subjected to a final degassing, after which nitrogen is admitted to the jar until the capsules are once more at atmospheric pressure. The vials are then stoppered quickly, leaving the capsules in a nitrogen atmosphere, after which they are polymerized with heat. This method certainly is unnecessarily complex for routine use, although it may be of value in particular situations.

4.5. "Polymerization Damage" and the Use of "Prepolymerized" Methacrylate

A principal variant method of methacrylate embedding involves using partially "prepolymerized" methacrylate instead of simple monomeres as the initial medium. This technique was developed by Borysko (1956 and subsequently) in an effort to avoid "polymerization damage" which may result from conventional procedures.

There since has been much discussion of polymerization damage. It is thought to result from nonuniform polymerization which disrupts the fine structure of some tissue and under some circumstances. Factors producing it may include shearing effects associated with the volume change (shrinkage) which occurs with polymerization, or which may accompany an advancing front of polymerization. The end effect can be hard, indeed, to distinguish from poor fixation, or from decomposition which may result when inadequately preserved material receives electron bombardment. (Morgan *et al.* (1956) have demonstrated the sensitivity of formalin fixed material to bombardment damage.) It is possible to prove that polymerization damage truly exists, however, by suitably controlled experiments. Samples of a single tissue embedded conventionally can be compared with samples embedded with prepolymerized methacrylate (Fig. 7), or after using epoxy or polyester embeddings.

Polymerization damage has often been likened to an "explosion." There is always some swelling. Note that the magnification of Fig. 7(a) is only half that of 7(b) so that the apparent size of the affected cell is about twice that of the well preserved one. Cell and nuclear membranes are apt to be interrupted. Mitochondria may be swollen and their membranes broken. Small vesicular structures, such as those associated with the Golgi apparatus and endoplasmic reticulum, may become almost unrecognizable. The diffuse proteins of the cytoplasm and nucleoplasm are apt to appear as a dispersed flocculent precipitate. It is usually obvious that the whole cell is poorly preserved. Polymerization damage can be a graded phenomenon so that varying degrees of explosion are recognizable.

It has been general experience that exposed cells on the surface of tissue blocks are the ones particularly prone to be damaged. Below the surface layer of cells, tissue ordinarily is unaffected by conventional embedding. Certain tissues are par-

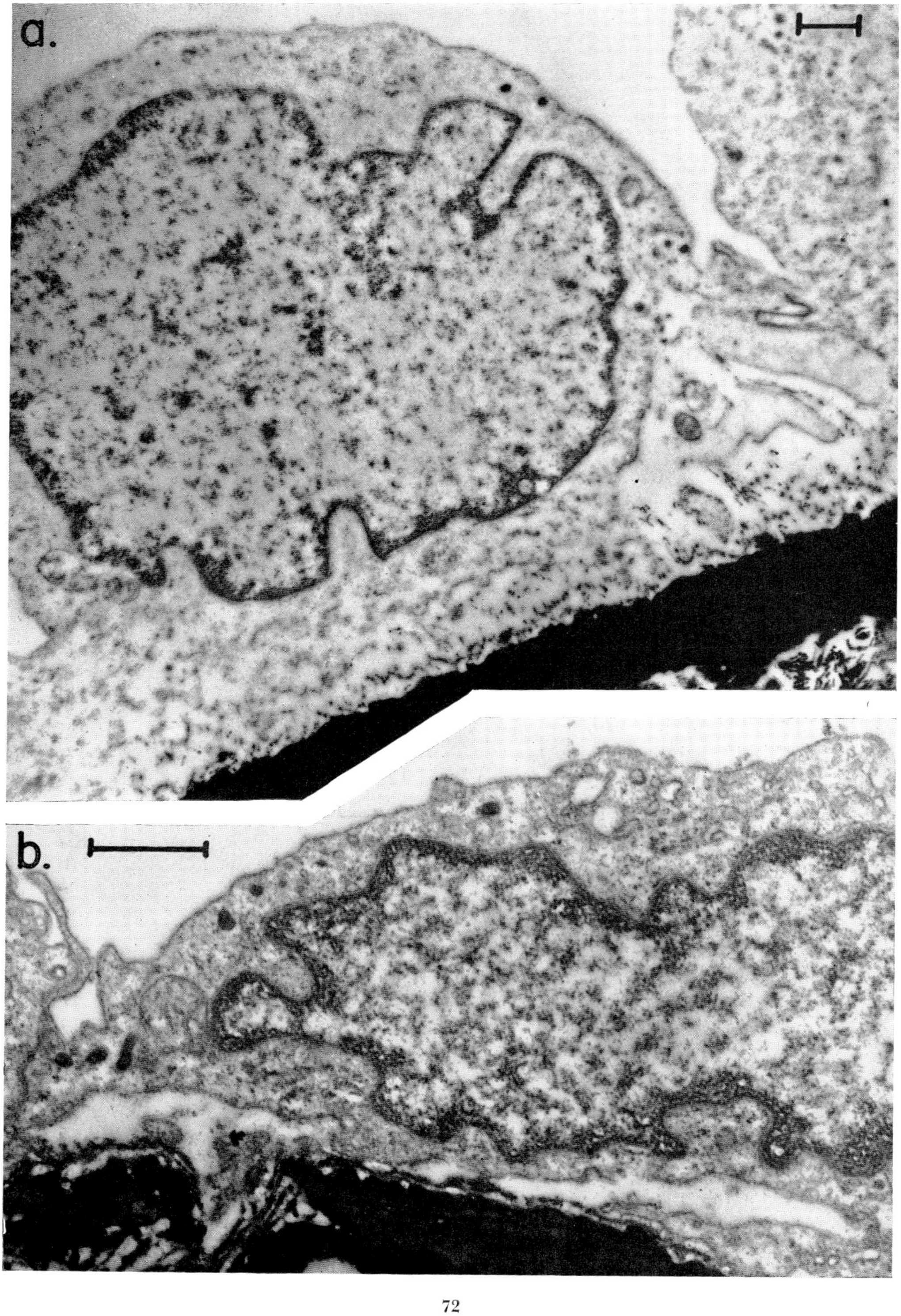
a.
b.

ticularly prone to damage for inexplicable reasons. Thus, in this writer's experience, the endothelium of slabs of an aortic wall is very sensitive (Fig. 7(a)). Adjacent muscle is not affected by conventional blocking, and exposed muscle on the opposite side of the aortic wall is unaffected even though the adventitial coat is stripped off prior to embedding. The endothelium of small blood vessels, blocked *in toto,* is not sensitive to damage. Thus, different cell types are unpredictably variable, and are affected only under particular conditions of exposure.

Polymerization damage has especially plagued investigators interested in tissue cultures; for in these preparations almost all of the interesting cells are on the surface of the blocks, and usually are sensitive to this sort of damage. It also has caused much difficulty when efforts have been made to handle free living cells such as protozoa, or free cells in suspension, as in blood or ascites fluid. In the latter instances the most successful microscopy has often been accomplished by centrifuging the cells to make a pellet, which then is handled as though it were a solid block of tissue.

The possibility of polymerization damage presumably is augmented by the degree of incompatibility of the different species of monomeres, which can be thought of as tending to polymerize with their own species at different rates. The damage can be expected to be most severe when relatively high concentrations of methylmethacrylate are used which start polymerizing before the butylmethacrylate. The damage theoretically should be less likely when the more compatible ethylmethacrylate is used rather than methylmethacrylate, and in particular, when its concentrations are kept minimal. This writer does not believe that polymerization damage is a problem when well-fixed material is embedded

FIG. 7. Endothelial cells from a single rat aorta. (b) This sample was embedded in "prepolymerized" methacrylate and showed uniformly a well-preserved endothelium. (a) This material was embedded conventionally with a mixture of methacrylate monomeres, and consistently showed serious "polymerization damage" in the *tunica intima.* This sort of damage suggests an "explosion." The cells appear swollen, in this case to about twice the size of the control. Membranes of all sorts are interrupted. Cytoplasmic and nuclear contents are dispersed, and have a flocculent appearance. This simulates poor fixation. Ordinarily, severe polymerization damage is limited to the surface of tissue blocks and freely exposed cells. Some cell types are more sensitive than others.

as described in the previous section, except in superficial layers of tissue blocks.

To avoid polymerization damage, Borysko (1956) advocated prepolymerizing methacrylate mixtures in bulk at 60–80° C with repeated stirring until they reach the consistency of a thick syrup. The reaction may take about a day at 60° C, and perhaps no more than an hour at 80° C. Since polymerization is an exothermic reaction, it may be hard to stop if high temperatures are used. The reaction must be watched carefully. Refraction patterns can be seen in the mixture as polymerization begins. If an Erlenmeyer flask is used for the polymerization, the reaction can be slowed relatively quickly by dipping in cold water. Also small quantities of cool monomere can be added safely to drop the temperature.

Prepolymerized syrup can be used at once by loading capsules and continuing polymerization at high temperature or with ultraviolet radiation, or the reaction can be stopped by lowering the temperature. Then the prepolymerized syrup can be stored for days or weeks in a refrigerator and used as needed. The latter procedure is regarded as a useful convenience in many laboratories for all embedding, quite apart from questions of specimen damage. But refrigerated containers must be allowed to warm almost to room temperature before opening them or water condensation may result. Large bubbles are apt to be formed and trapped in capsules containing prepolymerized methacrylate unless heat is applied to their bottoms at least as fast as to their tops. Therefore preheated racks should be used.

Prepolymerized methacrylate has the additional advantage that it can be adapted to an open-face, flat embedding (Borysko, 1958). In this way smears, spreads, hanging drop cultures, and the like can be embedded while attached to glass surfaces. After impregnating the material on a slide with monomere, aluminum foil is bent and folded around the slide to create a shallow tray. Prepolymerized plastic is then poured into this to a depth of 3 or 4 mm. The temperature is slowly raised to approximately 100° C which is maintained for 10–24 hr, after which it is slowly cooled (It would not be possible to use monomere in this way in uncovered containers for it would evaporate quickly.) Finally, the layer of plastic is stripped away from the glass carrying the specimen material with it. A variant method proposed by Howatson and Almeida (1958) inverts the lower half of a capsule containing pre-polymerized methacrylate over the desired cells. The capsule in contact with the glass is then put into the oven for hardening. If a thin glass surface such as a cover slip has been used, it can be detached from the methacrylate easily by placing it on a piece of solid carbon dioxide for a minute or two. They warn that it is then important to prevent water condensing on the cool block as this may cause swelling and distortion of the cells. They avoid condensation by transferring the blocks immediately to a desiccator and leaving them there until they have regained room temperature.

Efforts have been made to eliminate polymerization damage in various other ways. Recently Shipkey and Dalton (1959) have advocated the use of azodi-iso-butyronitrile as a nonoxidant catalyst for methacrylate embedding. They use it in concentrations varying from 0.5 to 1.5% with heat polymerization at 47° C (higher polymerization temperatures are apt to produce bubbles). They particularly recommend it for tissue culture material, blood cells, and cells on free surfaces where polymerization damage is a problem. Since this catalyst can be used to make prepolymerized syrup, one can perhaps obtain a double measure of protection. The authors believe it to be contra-indicated with solid tissues unless block sizes are kept very small; for they have had difficulty achieving uniform cutting qualities in the blocks.

Ward (1959) advocates the use of small quantities of uranyl nitrate added to the methacrylate monomere to prevent polymerization damage. He suggests using 0.75–0.37%. The effect is not entirely specific to the uranyl ion for there is some evidence that other heavy metals behave similarly. Unfortunately the cutting qualities of the blocks are altered by these additions. The polymerized plastic becomes harder than one would anticipate, and it is sometimes necessary to add a plasticizer. Dibutyl phthalate can serve in this respect, as much as 10% being added in some cases.

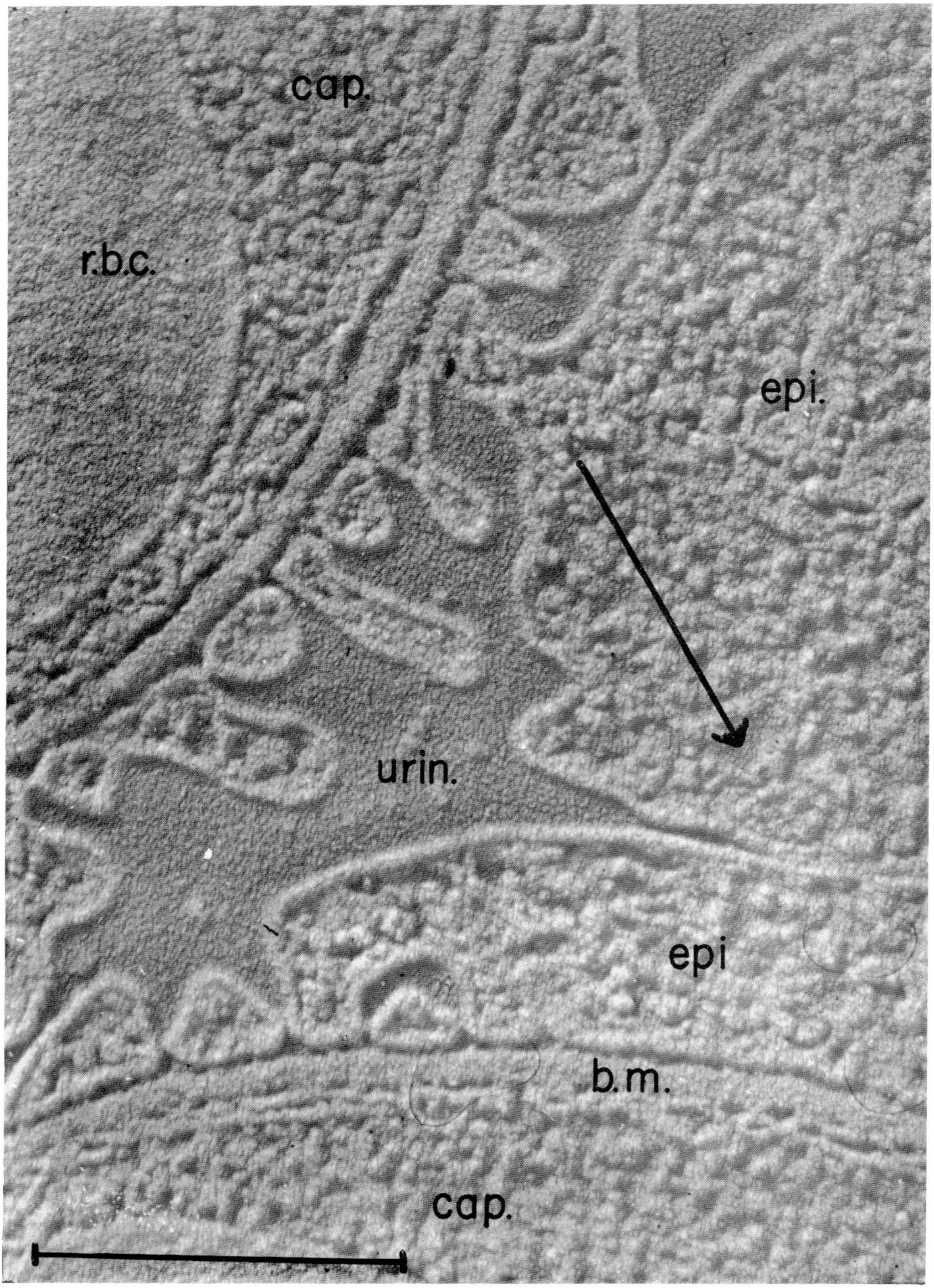
cap.
r.b.c.
epi.
urin.
epi
b.m.
cap.

The reader is referred to Gettner and Ornstein (1956) for a further account of polymerization damage and of efforts made to minimize it, and to Moore and Grimley (1957) for a theoretical account of polymerization which bears on this problem.

4.6. Alternative Embedding Media

In the introduction of this chapter allusion was made to an artifact associated with methacrylate sections which is apt to destroy detail of the finest resolvable order. One of the principal virtues of methacrylate is that it decomposes under electron bombardment so that much of its substance disappears leaving a background of relatively low electron density (Fig. 8). Unfortunately, as this occurs the methacrylate apparently melts, and surface tension forces come into play. Two membranes that are close together may fall upon each other so that subsequently they are observed as a single line. A cylinder may collapse. Two tiny granules that are close together may fuse into a single mass. Structures of complex angular geometry may be rounded. When one is primarily concerned with morphology at the macromolecular level these artifacts may indeed be serious (although to some extent "sandwiching" methacrylate sections between two supporting films protects them from this damage, see Chapter 6.10).

Examples from two other families of plastics have been developed as embedding media for ultrathin sectioning, and, in a large measure, eliminate the methacrylate artifact. The first of these historically, an epoxy resin "Araldite," decomposes little,

FIG. 8. This section of the wall of a glomerular capillary was exposed first to electron bombardment in an electron microscope, and subsequently obliquely shadowed with chromium in the direction of the arrow. It is apparent from the shadows that the surface was contoured, and that much of the supporting of methacrylate plastic had disappeared, leaving the specimen in relief, and thus subject to surface tension forces. Specimens embedded in Araldite or Vestopal W do not show this sort of sublimation. Sandwiching sections between two supporting films (see Chapter 6 and Figs. 21 and 30a) also is designed to minimize possible effects of surface tension.

This micrograph is printed as a negative as is commonly done with shadowed material so that the shadows appear dark. Legends are as follows: epi., epithelial cell; r.b.c., red blood cell; cap., capillary; b.m., basement membrane; urin., urinary space.

if at all, when bombarded by electrons at normal operating intensities of an electron microscope. Thus, macromolecular geometry presumably is unaffected by surface tension forces. The polyester, Vestopal W, apparently decomposes to some extent under electron bombardment, but sublimes in such a fashion that no liquid phase is created. Again there is little or no chance for surface tension to destroy fine structure.

An additional important reason for using Araldite or Vestopal is to avoid "polymerization damage" which seems to be a danger only with methacrylate embedding (discussed in the previous section of this chapter). Since material preserved with permanganate or formalin fixation is particularly prone to this, the investigator should consider using these newer plastics whenever he is using material preserved in some other way than with osmium tetroxide.

These important technical gains are not achieved without serious difficulties. In general, only those biological systems can be studied which can be "stained" to give them a contrast greater than osmium tetroxide fixation can provide (muscle, Huxley (1957)), or systems which are well preserved by permanganate fixation (nerve membranes and myelin, Robertson (1957)).

Finally, gelatin as an embedding medium deserves mention. Its advantage is that organic solvents are avoided entirely, so that substances are retained that might otherwise be lost. No doubt there are histochemical possibilities in the use of this method of embedding and sectioning, but these so far have not been exploited. These three embedding media are considered in the next sections of this chapter.

4.7. Epoxy Resin "Araldite" Embedding

Maaløe and Birch-Andersen (1956) first achieved partial success with an epoxy resin embedding medium. The particular resin they chose was not very satisfactory and subsequently Glauert *et al.* (1956) developed "Araldite" as a more effective embedding substance, although not without there remaining some difficulties in handling it. Glauert and Glauert (1958) have given an up-to-date account of its use as they currently employ it.

Their procedure begins with a conventional dehydration in ethyl alcohol. This is followed by a prolonged infiltration procedure which is essential for success since the large molecular species of the plastic, its "hardener" and its accelerator all diffuse slowly. Specimens, depending on their nature, are soaked for 1–6 hr. at a temperature of 48° C in a mixture of 2 parts absolute alcohol, 1 part Araldite and 1 part "hardener." They are then transferred to a pure mixture of Araldite and its hardener, and soft specimens soaked for 2 or 3 hr at the elevated temperature, while dense specimens may be soaked for as much as 24 hr. Specimens then are passed through two or three changes of the final mixture which consists of equal parts of Araldite and its hardener, to which 2½% of an "accelerator" and 5% of a "plasticizer" is added. Dense specimens are soaked in the final mixture for 24 hr at room temperature which does not initiate polymerization, and thus allows more time for equilibration. Polymerization is completed in two days at 48° C.

The Glauerts (1958) used Araldite M as their resin of choice, available from Aero Research Ltd., Duxford, Cambridge, and from Ciba Ltd., Basle. The European products are not sold in the United States. There is no exact American equivalent and, in fact, Araldite M is not a pure resin but one blended with a "modifier." The Ciba Products Corp., Kimberton, Pennsylvania state that the closest American equivalent would be a blend of 100 parts of Araldite 6010 with 24 parts of dibutyl phthalate. This is either identical with or similar to "Araldite 502" which is available in the United States as a commercial mixture.

The "hardener 964B" recommended by the Glauerts is dodecenyl succinic anhydride. The "plasticizer" is dibutyl phthalate. The "accelerator 964C," originally used by the Glauerts, was tridimethylaminomethylphenol. The hardener and plasticizer are commonly available, the accelerator is sold by Rohm and Haas Inc., Philadelphia.

J. Luft, in an unpublished study of epoxy embedding, has experienced the common difficulty of obtaining uniform impregnation, which leads to sectioning problems. He feels this can be largely eliminated by infiltrating first with two 30 min changes of propylene oxide. (This is a volatile and somewhat toxic substance, but since it is miscible with water, it can be flushed down a drain.) Then the blocks are soaked for 1–3 hr. in equal parts of propylene oxide and the Araldite mixture. The latter is made up of equal parts of Araldite M (or 502), and its hardener, plus 2% accelerator. Next the blocks are transferred to capsules containing the Araldite mixture avoiding, insofar as possible, carrying over the propylene oxide mixture. Polymerization is begun by heating the capsules overnight at 35° C. Then the temperature is raised to 45° C during the following day, and to 60° C for the final night when polymerization should be finished and the blocks ready for cutting. Birbeck and Mercer (1957) advocate passing tissue through xylol, and then xylol-Araldite-hardener mixtures to facilitate uniform infiltration.

D. Moore, in New York, has made a major effort to study this embedding process, and has analyzed the properties and effects of various possible hardeners and accelerators. His findings have not been officially published, although they are available in mimeograph form in a kit for Araldite embedding currently being sold by Cargille and Sons, Little Ferry, New Jersey. Moore advises handling the Araldite resin differently than other investigators in one important respect. He believes that the accelerator penetrates into tissue very slowly and, in effect, its penetration finally determines whether or not successful polymerization occurs within the tissue. He therefore recommends including the accelerator in the preliminary bath used for the impregnation of the tissue. He also recommends a different accelerator than that used by the Glauerts, *n*-benzyl dimethylamine, and the use of Araldite 6005 which is not a blended resin. Araldite 6005 was chosen because it had the lowest molecular weight of any unmodified resin available in the United States.

Moore begins impregnation with a bath of 5 parts absolute alcohol, 2 parts Araldite 6005, 2 parts hardener, and 1 part accelerator. Tissue is soaked in this for 3 hr. at a temperature of 48° C, with occasional stirring.*

The accelerator, except that which is within the tissue, must

* A more recent formulation reduces the amount of the accelerator to 1/10 of the Araldite.

be thoroughly eliminated from the next bath to prevent active polymerization. So, before proceeding, the tissue blocks are rolled about on filter paper, and are transferred to new vials. The second bath consists of a mixture of equal parts of Araldite and its hardener. Tissue is soaked in this for 24 hr. at 48° C with occasional stirring.

The third step returns the tissue to a solution containing accelerator, formulated with 20 parts of Araldite, 20 parts of hardener, and 1 part of accelerator. It is soaked in this for approximately 20 hr *at room temperature,* again with occasional stirring.

Finally, the tissue is placed in capsules which are then filled with this last mixture, and polymerized by overnight heating at 48° C.

Moore does not recommend using a plasticizer, but blocks prepared by his method are hard and excessively brittle, so brittle in fact that they cannot be trimmed without much fracturing and splintering. This writer has added 5% plasticizer to the final mixtures which at least makes the blocks easier to handle. Note, though, that the Glauerts' casting resin contains about 20% of a plasticizer to begin with, to which even more may be added. Thus, it is not surprising that Moore's blocks are very much harder and more brittle than, for example, J. D. Robertson's which this writer was privileged to examine in England.

The great viscosity of Araldite makes the preparation of the various mixtures somewhat difficult and uncertain. If one begins with a known amount of Araldite, however, other substances can be added in proportion. Furthermore, Araldite and its hardener can be stored in a 48° C oven which reduces the viscosity quite substantially. Similarly warmed graduates and pipettes should be used in mixing. It is convenient to keep separate pipettes for each of these materials, and simply drain and store them after use by standing them in their own 250-ml cylinders rather than trying to clean them each time.

In making up the Araldite mixtures it is most important that they be thoroughly blended. The thick Araldite is apt to layer on the bottom of a vial and on its sidewalls. It is not prone to mix easily with the hardener. Thus, it is necessary to stir vigorously

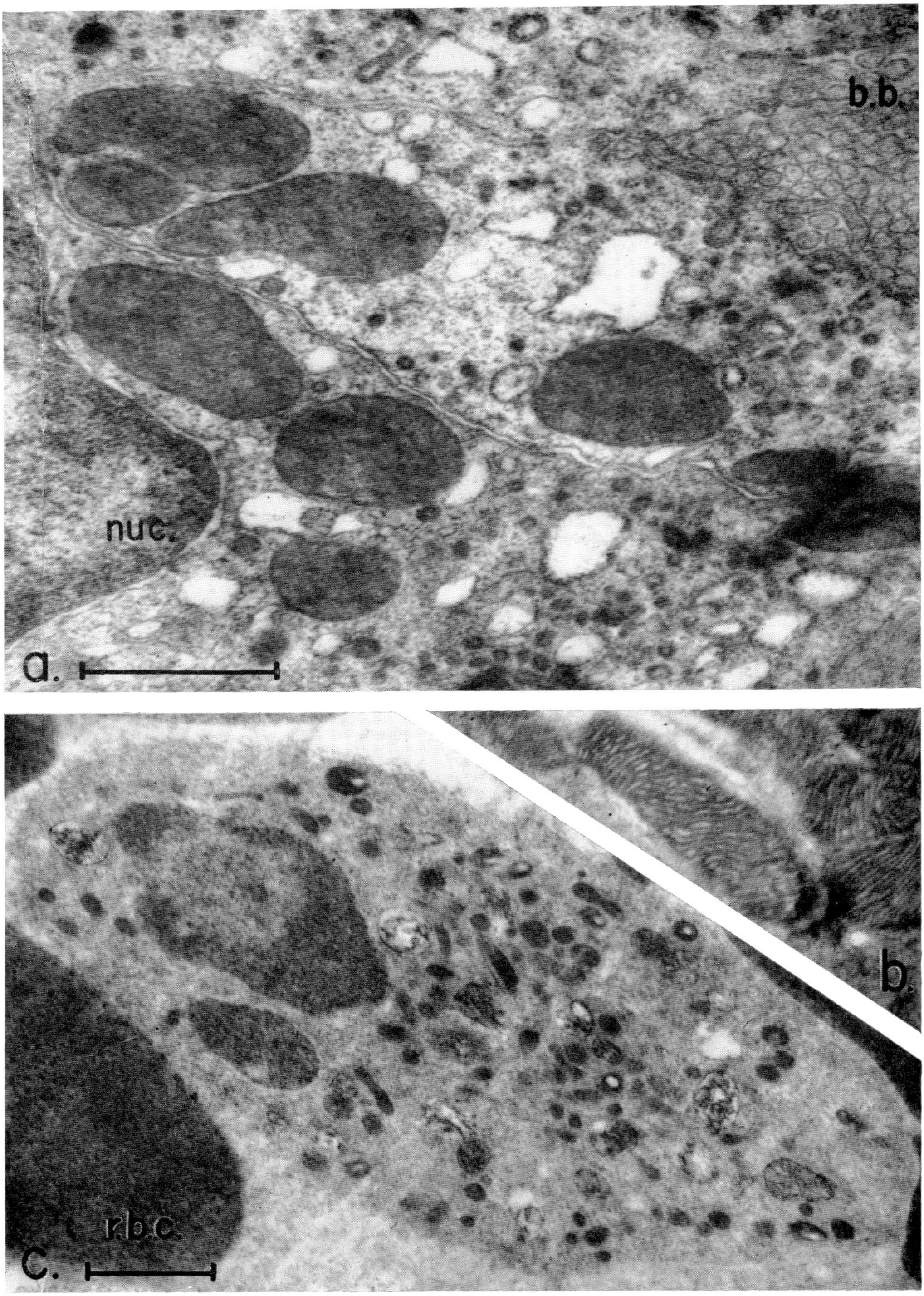
b.b.
nuc.
a.
b.
r.b.c.
c.

and for some time to be sure of the homogeneity of the mixture. An electric stirrer is almost mandatory for this.

Moore warns that some investigators develop a dermatitis from handling epoxy materials, and that both the vaporous and liquid forms of the accelerator are dangerous.

Between 1956 and the present, American laboratories have had great difficulties in achieving success with Araldite embedding. Perhaps this has been partly for want of sufficient personal contact with the English, and a precise knowledge of their exact methodology, but there is a tendency to blame the American made Araldite. This writer is inclined to agree with the latter charge from his own experience; for he has had opportunity to cut and examine sections from blocks prepared in England by J. D. Robertson. He has made an effort to compare these with blocks prepared in the manner of the Glauerts but with American materials, and with blocks made according to the method of Moore. Sections of suitable thickness (Fig. 9) have been obtained from blocks prepared in all of these ways in spite of textural differences. The actual sectioning of Araldite embedded material is not difficult, although cutting artifacts such as those of Fig. 9(c) have plagued this author's efforts.

The real difficulty Americans have had in attempting to use Araldite embedding is to see something in their sections when they are done. Specimen contrast has been so poor that the sections generally prove useless. This writer now is inclined to believe that there are genuine differences in the density of the plastic which account for the "fact" that English sections show substantially better contrast than American ones. This may be due to the presence of the "modifier," for when a modified American resin (Araldite 502) is used in the manner of the Glauerts it

---

FIG. 9. Material embedded in Araldite, stained before blocking with phosphotungstic acid in absolute alcohol. (a) The apical region of cells of the proximal tubule of rat kidney sectioned obliquely. Transverse sections of brush border processes (b.b.) are to be seen as well as part of a nucleus (nuc.), cell membranes, mitochondria, vacuoles, and RNA-granules. (b) Mitochondria, showing cristae, also from the rat proximal tubule. (c) A neutrophil and part of an erythrocyte (r.b.c.) in a capillary of rat kidney fixed *in situ*. The granular appearance, particularly of the erythrocyte, is due to coarse surface structure in the section, a common result of poor cutting using either Araldite or Vestopal W embedding.

seems to yield sections with better contrast than when unmodified resins are used as Moore advocates.

As this book is being published, this writer has learned of at least two independent attempts to develop "Epon" epoxy resins as embedding agents. They show promise of proving superior to and more reliable than the Araldite resins. Their relatively low molecular weight and consequent low viscosity makes them easier to handle, and no doubt aids consistent and uniform impregnation. Apparently it is not difficult to produce blocks with those desirable characteristics of softness, toughness, and elasticity such that relatively large sections can be cut smoothly and easily. They are said not to be prone to the "chatter" and "scrapping" artifacts that often are so bothersome with Araldite.

Detailed methods of handling these materials have not yet been published, and, indeed, must still be regarded as in developmental stages. Dr. J. Luft, however, has provided this writer with a tentative schedule currently in use at Seattle. Tissue, after standard dehydration, is passed through two changes of propylene oxide. It is then infiltrated for 1 hr. in equal parts of this solvent and mixed resin (with accelerator). Finally it is transferred to the polymerizing mixture in capsules. These are heated overnight at 35° C, at 45° C throughout the next day, and at 60° C during the second night. Cutting seems to improve with further aging, but this is not strictly necessary.

Luft controls the final hardness by blending two different resin mixtures. These are made up as stock solutions which can be kept for protracted periods in the refrigerator. They are appropriately blended just before use, when 1.5% of the accelerator DMP-30* also is added. It is essential that this be mixed in very thoroughly by stirring for at least 5 min. Mixture A, which alone would produce a relatively soft plastic, consists of "Epon 812," 62 ml, and DDSA,† 100 ml. Mixture B, capable of producing a substantially harder plastic, consists of "Epon 812,"

* DMP-30 (tridimethylaminomethylphenol), available from Rohm and Haas, Washington Sq., Philadelphia, Pennsylvania.

† DDSA (dodecenyl succinic anhydride), available from the National Aniline Division, Allied Chemical and Dye Corp., 40 Rector St., New York.

100 ml, and MNA,* 89 ml. A mixture of ⅔ A and ⅓ B is said to resemble butylmethacrylate blocks with about 10–15% methylmethacrylate added. This is the mixture which Luft himself prefers for most work. Reversing the proportions of A and B is rather like adding 20–30% methylmethacrylate to butylmethacrylate.

4.8. Polyester "Vestopal W" Embedding

In 1956 Kellenberger *et al.* demonstrated that polyesters might be used as embedding media. Early results admittedly were inconsistent, but by now (Ryter and Kellenberger, 1958) a better resin has been found than was originally used, and it seems that this can be employed reliably to give uniform blocks. The present resin of choice is Vestopal W. The W stands for the German word "weich," and contrasts with a Vestopal H, standing for "hart," which also is available.

Vestopal is not miscible with alcohol. Dehydration, therefore, is concluded with the tissue in anhydrous acetone. It is then passed through a graded series of Vestopal and acetone mixtures allowing 30–60 min in each. The first mixture is 1 part of Vestopal and 3 parts of acetone; the second mixture consists of equal parts of Vestopal and acetone; the third mixture is compounded of 3 parts of Vestopal and 1 of acetone. The tissue is then ready for final infiltration with pure Vestopal plus 1% of an initiator and 0.5% of an activator. Specimens are left at room temperature in this mixture for 30–60 min, and then transferred to capsules with fresh mixture. The capsules are polymerized at 60° C overnight or for 24 hr.

Vestopal is made by the Chemische Werke, Hüls, but is available in small quantities along with the activator and initiator from Martin M. Jaeger, Vésenaz, Geneva, Switzerland. Ryter

* MNA (methyl nadic anhydride), also available from the National Aniline Division, Allied Chemical and Dye Corp.

Epon 812 is available from the Shell Chemical Corp., Shell Bldg., San Francisco, California.

The investigator who is interested in general information about Epon resins will find a good summary in *Technical Bulletin*, SC:58-88, of the Shell Chemical Corp.

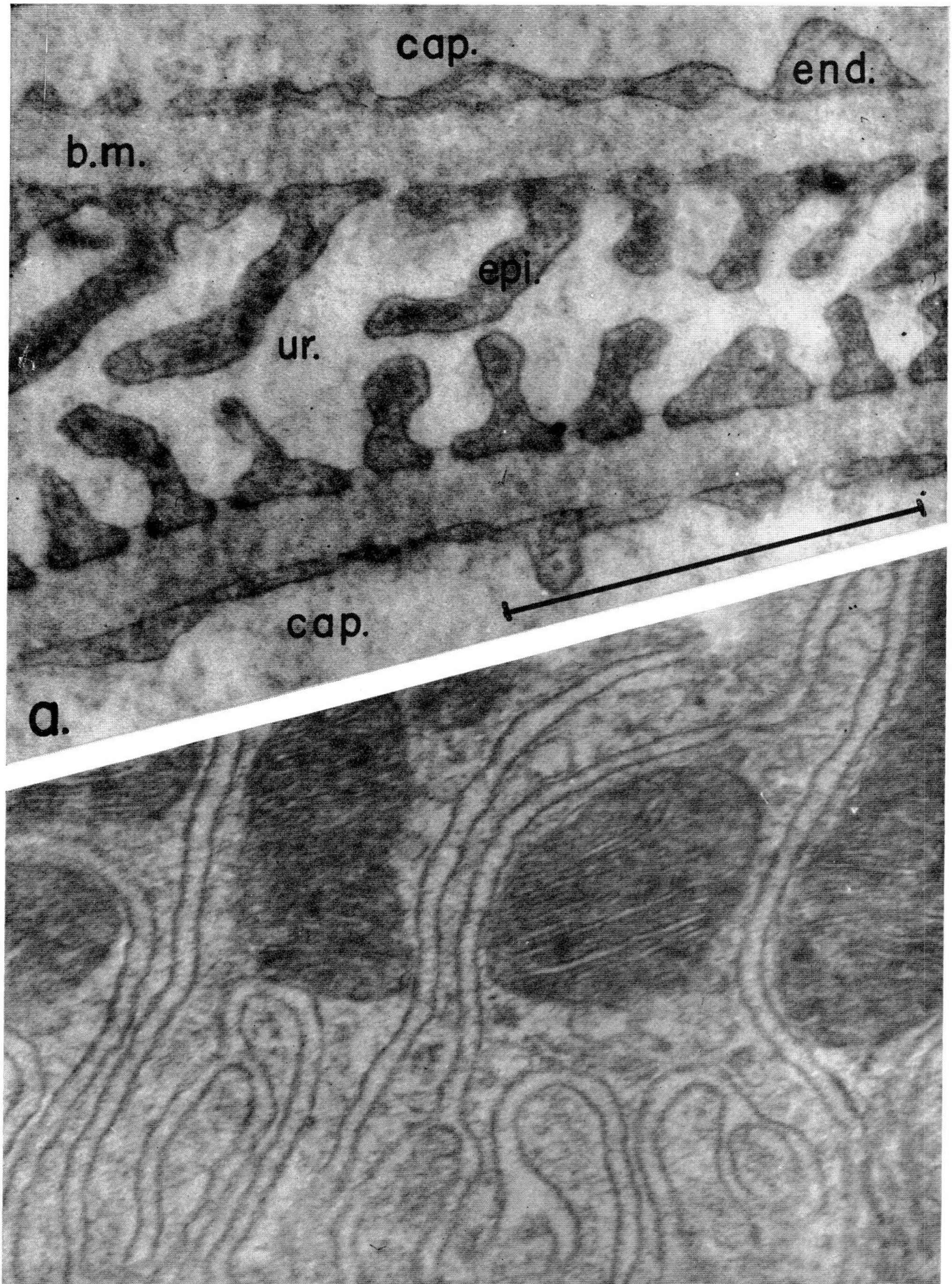
cap.
end.
b.m.
epi.
ur.
cap.
a.
b.

and Kellenberger prefer tertiary butyl perbenzoate as the initiator, and cobalt naphthenate as the activator. They warn that these substances must not be mixed directly together for the combination is explosive. These substances are said not to keep well, even in a refrigerator and protected from light, and must be replaced every few months. This is an unfortunate combination of properties for Americans since it may complicate getting the materials to this country from Switzerland. Substitutes may prove possible. Ryter and Kellenberger indicate that benzoyl peroxide will serve as an initiator, while lauryl mercaptan can be employed as an activator. The difficulty with benzoyl peroxide is that it is solid and difficult to dissolve in Vestopal. One can imagine that the objection to lauryl mercaptan is principally one of odor.

In a pamphlet distributed by the Martin M. Jaeger Company the following difficulties are anticipated. If the activator becomes thick or changes color it should be discarded. If cracks appear in the blocks either the initiator or the activator was not mixed well enough, or the polymerization was too rapid from too high temperature. If the tissue is shrunken, the passages from acetone into plastic were made too quickly. If the material in the section has a spongy appearance, acetone was not entirely removed from the tissue. Mixtures containing resin with initiator and activator will not keep and must be made up freshly before each use.

In this writer's experience, Vestopal embedded blocks are very much harder than we are accustomed to in this country. Indeed, they are so hard and tough that they are difficult to trim

---

FIG. 10. Rat kidney material embedded in Vestopal W. (a) Parts of two glomerular capillary walls. Epithelial feet (epi.), basement membranes (b.m.), and endothelium (end.) are visible, as are also capillary lumina (cap.) and the urinary space (ur.). (b) The basal portion of a cell of the distal tubule showing deep infoldings of the plasma membrane, and mitochondria with indications of cristae.

Since this embedding material does not sublime as much as methacrylate, specimen contrast inevitably is much lower. This specimen had been "stained" before embedding by prolonged soaking in phosphotungstic acid, and also uranium acetate dissolved in absolute alcohol.

with a razor blade. Dibutyl phthalate may be used as a plasticizer, however, in concentrations of from 2 to 10%.

Sections of Vestopal embedded material are illustrated in Fig. 10.

4.9. Gelatin Embedding

A general technique of gelatin embedding has been developed by Gilëv, which was published in detail in 1958. Apparently independently, Fernández-Morán and Finean (1957) have also successfully experimented with a similar technique to study nerve myelin by low angle X-ray diffraction, as well as by electron microscopy. Since the present author has not had experience with this embedding medium he will only attempt a short summary here, and refers the interested reader to the original literature for further details.

Residual, uncombined fixative should be fully washed out of the tissue before beginning impregnation. Even after extensive washing Gilëv reports there is some interaction between osmium tetroxide and the embedding gelatin.

Next, the tissue is infiltrated with successively stronger gelatin solutions. Fernández-Morán and Finean used a commercially available, standard bacteriological gelatin; Gilëv purified a food gelatin by somewhat elaborate means. The former workers used relatively short time intervals to complete infiltration with a 30% solution to which 2% glycerin had been added. The latter worker used prolonged intervals to achieve infiltration of a 40% solution. Temperatures were maintained at 40° C or 37° C, respectively.

Following gelatin impregnation, Fernández-Morán and Finean solidified their gelatin by chilling it after pouring it into a Petri dish, and then they cut out the blocks they wanted. These were next placed in a vacuum desiccator and dried for 5–6 hr, the pressure finally reaching 10^{-3} mm Hg. The blocks so obtained apparently were very hard and difficult to section.

Gilëv used only air drying at 37° C to harden his blocks. This was deliberately done slowly by keeping a moist atmosphere. He describes his final blocks as "hard but not brittle." He regards as desirable some traces of water in the blocks at the time of sectioning, yet there must not be too much humidity. He flattened his sections by floating them on water, or a 0.5% solution of potassium-chrome-alum (subsequently washed out), or even 2% acetic acid. These treatments undoubtedly removed much of the gelatin, and hence the support, from the specimen, and so must be regarded as a dangerous and probably undesirable step.

4.10. Preparing the Block for Sectioning

For some time to come methacrylate certainly will remain the common embedding material for ultrathin sectioning, so that the following remarks will be concerned particularly with what to do with a hardened capsule of this plastic containing a specimen in one end of it.

If the specimen need not be cut with any particular orientation, the end of the capsule is simply trimmed as a truncated pyramid. The face which is to be sectioned should be trimmed actually into the tissue mass. Nothing is more frustrating than to prepare sections and discover when they reach the microscope that they are of empty methacrylate. In trimming freehand under a binocular dissecting microscope one ordinarily can see osmium fixed material in the relatively thick sections that are cut in facing the block. If there is doubt about whether tissue has been reached, the freehand sections can be transferred to a glass slide, then covered with a drop of immersion oil, and examined with a compound microscope using the condenser diaphragm nearly closed. Enough contrast usually can be obtained to see embedded tissue even though it is not stained. Better yet, one can use phase microscopy.

After facing the block the side walls are trimmed to make a square or oblong face that is small indeed by standards of light microscopy. The width of the block face to be cut ordinarily should be about 0.5 mm wide. If the material is "difficult" the block face may have to be substantially narrower even than this. (The most generally "difficult" material is that containing dense bundles of collagen. Some secretory products and, of course, mineralized tissue also may cause trouble.) Under the most favorable of circumstances the block face should not exceed 1 mm in width. Sections from wider blocks are almost certain to show "chatter" effects (Fig. 17).

The length of the block face is not as important as its width, although extreme lengths must be discouraged. In general, the length should not exceed twice the width.

It is permissible, indeed desirable, to trim the side walls of the block into the tissue mass itself. Unlike in conventional microtomy there is no advantage in having empty methacrylate surrounding the tissue in a section. All that a rim of blank methacrylate does is to increase the hazards of sectioning by making the sections unnecessarily large, and to cause irritation when finally examining the sections in the electron microscope.

In this writer's laboratory it has been a standard practice for several years to give the specimen added support with a superficial layer of hard wax. This has two further advantages. If the knife hits wax before methacrylate on the leading edge of the block, the knife gets started well. On the trailing edge of the block the final threads of methacrylate are not always cut easily unless a wax layer provides a backing that assists this process. So, after final trimming, we routinely drop our blocks into a melted and filtered mixture consisting of 1 part carnauba wax and 2 or 3 parts of paraffin wax. To keep the carnauba wax melted, the oven must be maintained at over 80° C. The blocks are exposed to this bath for approximately 3 min. This is sufficient time so that there is some exchange of wax and methacrylate at the surface of the block. Subsequently, crystals of wax can be seen within the most superficial layer of the plastic. Thus, there is an excellent bond established between the plastic and the wax. A suitably thin coat of wax will be obtained if the block is drained as much as possible while it is still hot. If necessary a mildly heated scalpel blade can be used for moving wax about. If a knife is used, however, care must be taken to be sure that the blade is clean so that grit is not introduced into the wax layer.

When it becomes important to orient a tissue specimen in some special manner, it often becomes necessary to mount the block on some sort of a holder. Short sections of wooden dowel rods, 5/16 in. in diameter, will fit into the chuck of a Porter–Blum microtome and serve this purpose very well. The dowel rod can be trimmed with a hand pencil sharpener as indicated in Fig. 11(e). The end of the dowel which is to receive the specimen block is prepared by dipping this in the melted carnauba–paraffin wax mixture indicated above.

If the orientation of the specimen must be changed considerably from the way it lies in the capsule, it is convenient to trim the specimen block almost to its final form while it is still left attached to the main mass of methacrylate. The latter serves as a handle making the operation easy to complete if the necessary razor cuts are planned in advance so they can be done in a reasonable sequence, always leaving plastic to push against. Figures 11(a)–(f) indicate how a flat piece of tissue can be trimmed so that finally it is mounted on edge to be cut transversely.

A flat piece of tissue, of course, sinks to the bottom of the capsule as in Fig. 11(a) where it can only be cut in the horizontal plane if it is not removed from the capsule and remounted. In Fig. 11(c) the truncated pyramid remains attached to the main mass of the capsule by only a narrow neck. At this stage it is immersed briefly in the molten carnauba–paraffin mixture to allow good bonding. The neck of methacrylate then is cut without difficulty, and the block is oriented on the dowel rod as indicated in Fig. 11(e). It usually is necessary to melt the wax and maneuver the specimen with a hot scalpel blade to accomplish this. This, of course, is done with the aid of a low power binocular dissecting microscope. When the orientation is regarded as satisfactory, the tip of the dowel including the block can be dipped in the hot wax bath to build up the wax mass if necessary (Fig. 11(f)). It is held in the hot wax long enough so that good fusion occurs. Finally, excess wax is maneuvered away from the cutting area with a warm knife blade, being most careful to avoid contamination of the wax with particulate matter which might subsequently damage the knife edge. Obviously, one must be judicious in the use of a hot scalpel blade for these operations, for too hot a blade naturally can damage the specimen.

If it is only necessary to make a simple or approximate reorien-

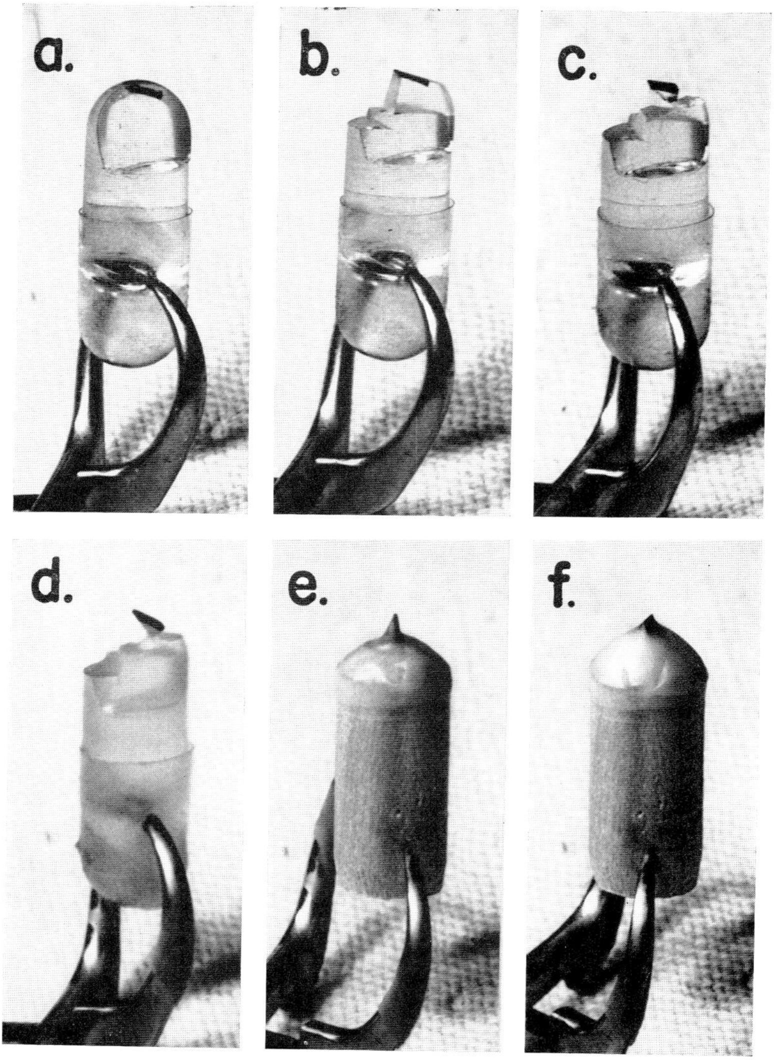

FIG. 11. An unoriented specimen embedded in methacrylate in a capsule can be oriented at will by mounting it in wax on a short segment of wooden dowel-rod. To obtain good bonding with the wax the specimen first should be trimmed almost completely, (a)–(c), and then be exposed to the melted wax for about 3 min during which time an interchange of wax and plastic occurs near the surface (d). Then it is oriented on the dowel-rod in a puddle of wax and teased into the

tation of the tissue, the piece of tissue can be given a flat face which will become its base, and this be stuck down to the squared-off end of a methacrylate plug simply by softening the surfaces to be joined with chloroform. Alternatively it can be glued with "household cement."

Variations of these basic procedures to meet particular situations will, in general, be obvious. If one wishes a hard wax with a lower melting point than carnauba, bayberry wax can be substituted. However, although bayberry wax is hard, it is extremely brittle, while carnauba is quite tough. Neither of these waxes, even when combined with paraffin, adhere well to most surfaces unless they penetrate. If a sticky wax is needed, beeswax can be added to the mixture, although this is relatively soft. All of these waxes are apt to be contaminated with gritty material when they are purchased and must be filtered before use.

Literature Cited

Birbeck, M. S. C., and Mercer, E. H. (1957). *J. Roy. Microscop. Soc.* **76,** 159.

Borysko, E. (1956). *J. Biophys. Bichem. Cytol.* **2** (Suppl.), p. 3.

Borysko, E. (1958). *J. Appl. Phys.* **29,** 1620.

Fernández-Morán, H., and Finean, J. B. (1957). *J. Biophys. Biochem. Cytol.* **3,** 725.

Gettner, M. E., and Ornstein, L. (1956). *In* "Physical Techniques in Biological Research" (G. Oster and A. W. Pollister, eds.), Vol. III, Chapter 13. Academic Press, New York.

Gilëv, V. P. (1958). *J. Ultrastructure Research* **1,** 349.

Glauert, A. M., and Glauert, R. H. (1958). *J. Biophys. Biochem. Cytol.* **4,** 191.

Glauert, A. M., Rogers, G. E., and Glauert, R. H. (1956). *Nature* **178,** 803.

Howatson, A. F., and Almeida, J. B. (1958). *J. Biophys. Biochem. Cytol.* **4,** 115.

Huxley, H. E. (1957). *J. Appl. Phys.* **28,** 1370.

---

proper position (e) and given the maximum possible support by building up wax around it (f).

The specimen should be so trimmed that nothing but facing is required of the finished mount (f). To do this conveniently, the trimming process should be plotted in advance, and the sequence such that as much support as possible is retained during the cutting. Usually it is best to leave the specimen attached to the capsular mass through the wax dip, for then one has a convenient "handle." Finally the specimen is detached by cutting the last remaining thin neck of plastic. The geometry of the specimen and the orientation required will determine just how these operations are performed, but a flat sheet or a long cord of tissue can easily be mounted vertically in this way irrespective of how it happened to lie in the capsule.

Kellenberger, E., Schwab, W., and Ryter, A. (1956). *Experientia* **12,** 421.

Maaløe, O., and Birch-Andersen, A. (1956). Bacterial Anatomy. *Symposium Soc. Gen. Microbiol.* **6,** 261.

Moore, D. H., and Grimley, P. M. (1957). *J. Biophys. Biochem. Cytol.* **3,** 255.

Morgan, C., Moore, D. H., and Rose, H. M. (1956). *J. Biophys. Biochem. Cytol.* **2** (Suppl.), p. 21.

Newman, J. B., Borysko, E., and Swerdlow, M. (1949). *J. Research Natl. Bur. Standards* **43,** 183.

Pease, D. C., and Baker, R. F. (1948). *Proc. Soc. Exptl. Biol. Med.* **67,** 470.

Robertson, J. D. (1957). *J. Biophys. Biochem. Cytol.* **3,** 1043.

Ryter, A., and Kellenberger, E. (1958). *J. Ultrastructure Research* **2,** 200.

Shipkey, F. H., and Dalton, A. J. (1959). Proc. Elect. Micro. Soc. Amer., *J. Appl. Phys.,* **30.**

Ward, R. T. (1959). *J. Appl. Phys.*

5. Sectioning

5.1. Introductory Remarks

Ultrathin sectioning has been possible for just over a decade. In retrospect, of course, the early beginnings were relatively crude.

Attempts at ultrathin sectioning actually date back to the pre-war years and von Ardenne's (1939) effort to cut tapering wedges which would be satisfactory in at least part of their area. Later, Richards and associates (1942) and Sjöstrand (1943) continued this approach but with little success.

O'Brien and McKinley (1943) proposed that very high cutting speeds were necessary if ultrathin sections were to be obtained, and they made an effort to accomplish this by, in essence, attaching a razor blade to a centrifuge rotor. Fullam and Gessler (1946) made a more serious effort to design an instrument to operate on this principle, and to use very high speeds. Such a microtome was actually marketed two years later with the advertisements suggesting that the investigator could protect his investment by easily converting his microtome to an ultracentrifuge.

Pease and Baker (1948) showed that ultrathin sectioning at reasonable cutting speeds was possible if a suitable embedding medium was used. They were also able to modify a conventional microtome for this purpose.

Newman *et al.* (1949) introduced butylmethacrylate which still remains the embedding medium of choice. These investigators also first employed a thermal expansion principle to drive a microtome.

The next few years saw a variety of microtomes being developed in different laboratories. Some of these were modifications of conventional instruments, others were original designs; a few were commercially marketed. Although it was apparent that these microtomes could be made to work in the hands of their designers, in general they were not sufficiently reliable to produce satisfactory sections in inexperienced hands.

Microtomy was aided greatly when sections were floated away from the knife edge on fluid surfaces, an idea perhaps first employed by Claude (1948), but exploited by Gettner and Hillier (1950).

At first, steel knives were used for microtomy and considerable ingenuity was employed in perfecting their edges. In spite of this there were limitations and difficulties, and the discovery by Latta and Hartmann (1950) that fracture edges of glass could be used was a much appreciated boon.

The years between 1948 and 1952 thus were a time of ferment. Bretschneider (1952) could review the history of ultrathin microtomy and still discuss virtually every paper that dealt with tissue sections. Today, not one of these papers is of more than historical interest. But for those of us who experienced this period it was a time of intense interest. We were confident that we could break through the many technical difficulties that plagued us, for it was apparent that these were fundamentally of a minor nature.

The definitive years were 1952 and 1953 when the first micrographs that we would still regard as of good quality appeared from the Rockefeller and Karolinska Institutes. Yet it is not really fair or possible to attribute their success to any one advance. Rather, it represented the culmination of improvements in microtomy, as well as a better understanding of problems of fixation, and also instrumental developments in the electron microscope itself such as the biased gun, compensated pole-pieces, and the use of apertures to increase specimen contrast.

An important event for the development of ultrathin microtomy was the appearance of the commercial version of the Porter–Blum microtome. This was first described in 1953 and became generally available that year. This microtome has clearly become the standard American instrument. The ultramicrotome developed by Sjöstrand (1953) which became available commercially about the same time enjoys a somewhat comparable position in Europe. It seems that proximity to the source has played a large role in deciding the choice. These microtomes had sufficient reliability so that many laboratories could begin effective work. The introduction of these instruments completed the technical "breakthrough," and the electron microscopy of tissue sections has prospered ever since.

We are currently seing a second wave of interest in microtome design. This is perhaps particularly true of Europe rather than of America. At the Fourth International Congress of Electron Microscopy in Berlin (1958) more than half a dozen new instruments were shown by commercial exhibitors as being either on the market or available within months. No doubt this recent interest reflects a degree of dissatisfaction with present techniques of microtomy and the hope that more automatic and more flexible machines can be developed. However, section quality is not limited by the mechanics of existing microtomes, and these new instruments can have little to offer other than features of convenience and automation.

The reader interested in further information about the early development of microtomes and microtomy should consult the excellent review of Gettner and Ornstein (1956).

5.2. The Porter-Blum Microtome*

This writer must speak mainly about the Porter–Blum microtome with which he has had almost all of his recent experience. It probably will be the instrument bought by most American laboratories for some years to come. Its mechanical advance is a feature not generally incorporated in the newer microtomes, yet this is of great value in practical microtomy. Almost all other microtomes which have been built have relied upon thermal expansion to advance the block. This sounds good in theory, but is wretched in practice, for it destroys operational flexibility in ways that will become apparent later. If it were not possible to make a good mechanical drive one might be willing to fall back upon thermal expansion as a necessary alternative. How-

* Manufactured by Ivan Sorvall Inc., Norwalk, Connecticut.

ever, the proven reliability of the Porter–Blum microtome demonstrates that a mechanical system of screw thread, lever arm, and suitably designed bearings is possible and practical. Those who object that section thickness must then always be a multiple of some unit can, if they insist, use the Porter–Blum microtome with thermal expansion drive. The prototype model, in fact, lacked a mechanical drive, and advance was by moving a lamp bulb close to the arm which carries the specimen.

If one insists upon automation the Porter-Blum microtome can be equipped with a motor drive. However, the microtome is quite sensitive to high frequency vibration which is difficult to damp out sufficiently. Also, a simple motor drive means that the microtome cycles at a uniform rate of speed. Since the actual cutting speed has to be slow, the whole cycle is then unnecessarily long, and thermal fluctuations become a serious problem. (This dilemma is not solved by most other microtomes which are motor driven, and remains an inherent weakness. Hand operation allows the cutting speed to be varied at will while the cycling remains relatively fast.)

The Porter–Blum microtome is capable of prolonged operation without maintenance. In fact, the manufacturer advises not making any attempt to lubricate the all-important pivot bearings, and advises only an occasional light oiling of the ratchet wheel and its associated moving parts, and the spindle of the drive wheel. Only very thin oil should be used. They provide a special lubricant which may be used on the guide plate bearing which controls the movement of the arm.

In our own laboratory we have had pivot bearings "freeze" twice. Although the manufacturer recommended the return of the instrument to the factory for repair, we did not encounter any particular difficulty in replacing the bronze springs and ball-bearings ourselves. The only other difficulty we have encountered is in having the pawl mechanism bind on the "limiting disk" so that the tooth did not regularly contact the ratchet wheel and advance the mechanism. If there is doubt about the advance, one can feel from below whether the main screw is turning as the microtome is cycled. Cleaning and lubricating the pawl mechanism should be all that is required if sticking occurs.

5.3 Preparation of the Porter–Blum Microtome

In this writer's opinion the Porter–Blum microtome should be modified in two minor ways to make its operation more effective.

It seems desirable to put a drag on the microtome which otherwise turns so freely that it is hard to control the speed very precisely. A simple and easily adjustable drag can be made if the fly-wheel is mounted in a lathe and a groove (with a V-shaped cross section) cut in its rim. A piece of heavy but flexible cord, or leather belting, which is greased and fastened to the table at one end, can be dangled over this to provide the drag. The amount of the drag depends upon the length of the dangling cord, or upon weights which may be attached. Just enough drag should be established so that one has something to push against in cycling the instrument. This modification is partly visible in the lower right corner of Fig. 16.

The other modification involves turning the fly-wheel and handle on its shaft so that the handle approaches the top of its turn just about at the same time as the block is cut. Thus, the operator's hand pushes forward against a slight pressure at the moment of sectioning. It seems to this author that this motion is the steadiest and most controllable of the whole cycle of the hand, although other workers might find that pulling the handle towards them at the bottom of the turn as sections are cut might be an equally good or preferred position. In any case, at the critical moment of sectioning, the hand should be moving in a horizontal plane rather than a vertical one.

Some thought must be given to where the microtome is located. It is sensitive to fairly high frequency vibration such as is produced by electric motors, but is not particularly sensitive to a random jolt. It is probably best to set up the microtome on its own small but sturdy table, which must have all four feet resting firmly on the floor. If there is any question of walls transmitting vibration, the table should stand free of them. No particular damping measures are necessary if the floor is reasonably stable. Obviously, drafts should be avoided where possible, but the microtome is surprisingly insensitive to thermal fluctuations if it is cycled by hand at a reasonable rate. Thus, elaborate temperature control is unnecessary.

5.4. Preparation of the Sjöstrand Microtome*

This writer has had no experience with the Sjöstrand microtome beyond a demonstration and talks with Dr. Sjöstrand, and so he cannot speak with authority about its operation. It is motor-driven, and apparently it is sensitive to vibration from this source. Thus, in Sjöstrand's own laboratory the microtome is mounted on a pedestal of poured concrete, and the motor is mounted on a completely separate concrete pedestal about 2 ft away. Connection between the microtome and motor is by a long and flexible belt. The microtome does not seem unduly sensitive to the transient jars that inevitably accompany the removal of sections while the instrument is operating.

5.5. Knives Used for Ultrathin Sectioning

Until very recently, ultrathin sectioning was being done almost exclusively with glass knives. These are made to be used once and then discarded. They need not cost anything.

Now diamond knives, originally developed by Fernández-Morán (1956), are becoming commercially available. Their cost is considerable, but they last indefinitely (although not without resharpening after each few months of use). Tissue sections cut with a diamond knife are no better or worse than ones cut with glass; so the choice is one of expense and convenience. Some knowledge and skill is required to prepare glass knives which is the topic of the next section.

* Manufactured by the LKB-Produkter, Fabriksaktiebolag, Stockholm.

5.6. Glass Knives

Failure in ultrathin sectioning can easily be due to the knife, yet this sort of failure is completely unnecessary. When the preparation and handling of the knife is properly understood, difficulties from this source cease to be a problem.

Above all else, the edge of the glass knife must be "clean" if it is to be used successfully for ultrathin sectioning. This cannot be stressed too strongly. Without any question the best way of obtaining clean edges is to prepare them just before use. No doubt it is tempting to the novice to purchase knives already prepared by professionals, but these knives inevitably are contaminated, and the novice is in serious trouble right at the start without his realizing it. In this laboratory we have not tried to clean an old knife for years, having long ago given it up as an unreliable and usually futile task.

It is so very easy to prepare one's own knives that, for this reason alone, it is foolish to buy them. Almost any type of glass of suitable thickness will make satisfactory knives. Easiest to get, and certainly cheapest, is the ordinary "crystal" glass, 3/16 in. thick, which is used for store windows of moderate size. The larger glazing companies accumulate barrels full of edge trimmings of such glass which are free for the asking. One selects strips 1–2 in. wide from which knives can be made simply by breaking them diagonally (Fig. 13). Companies which specialize in manufacturing mirrors will also have satisfactory edge trimmings of "plate" glass of suitable thickness. This is equally good with or without a mirrored face. Of course, it is also possible to have large sheets of glass commercially broken into strips. The most convenient width is approximately 1½ in.

Some selection of the glass strip used to prepare knives is necessary. One edge of the strip is finally going to be the vertical face of the finished knife. This should not be too severely checked (as is the case in Fig. 12(e)), and, in any event, checkmarks must not intersect the final edge. Additionally it should not be very wavy (as in Fig. 12(d)) for protruding bumps on the face of the knife might hit and compress the block face. Also, if the front

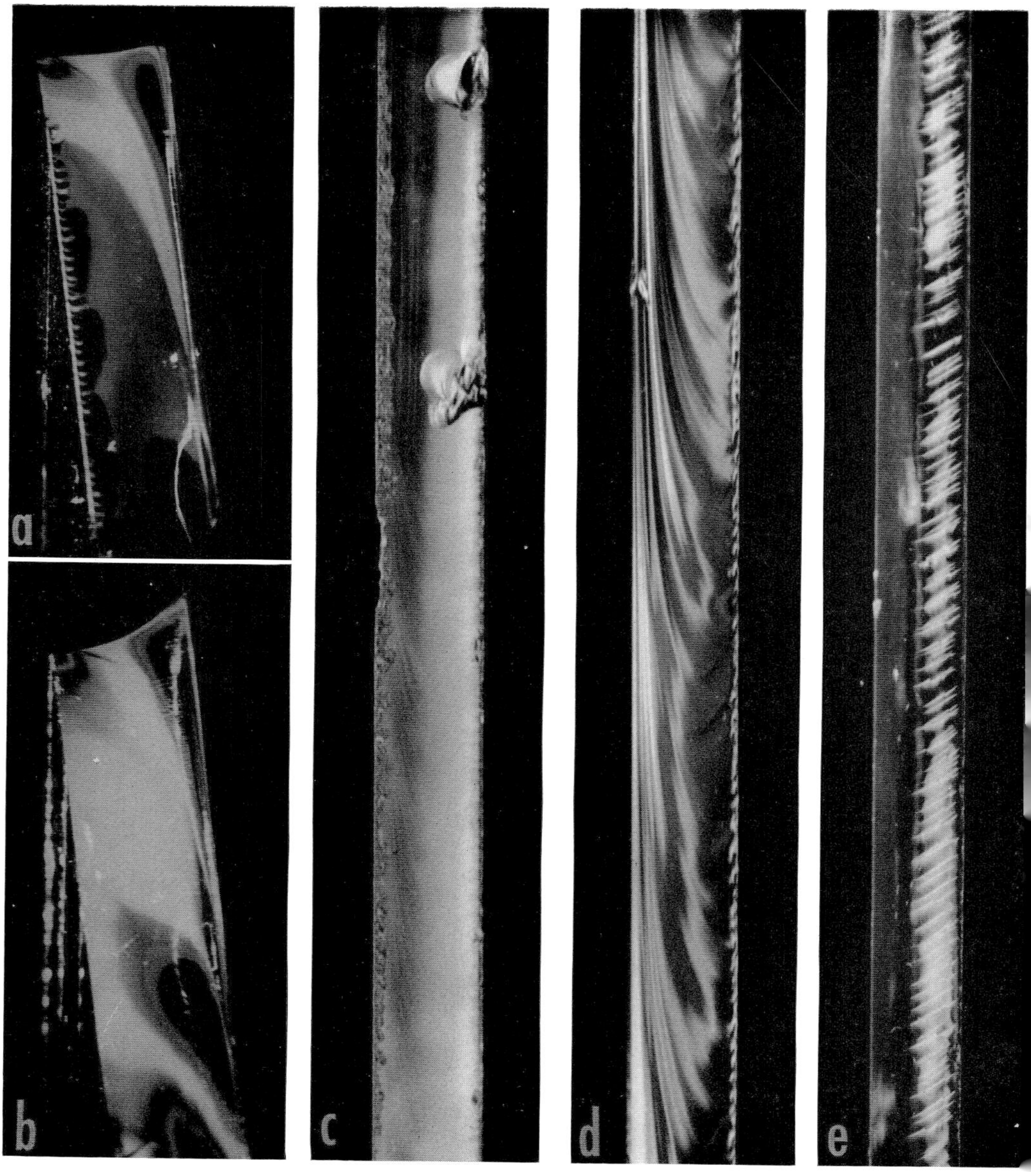

FIG. 12. (a) A good glass knife viewed from the top. Note the nearly horizontal edge. (b) A poor knife with a high spur on the right, and a severely sloping edge.

face of the knife is irregular, one has difficulty deciding what angles exist at the edge itself, and, consequently, it is not possible to judge the tilt of the knife accurately. One can see these defects by sighting obliquely along the edge of the glass strip and observing the patterns of light which reflect from its surface. A strip of glass with desirable properties (as in Fig. 12(c)) should be reasonably smooth and have few if any checks, except at the edge where the cutting tool inevitably produces some flaking.

The edge of the strip of glass which finally will be the front face of the knife must be meticulously cleaned before the knife is broken. It is essential that it not be contaminated subsequently. Fingers, particularly, have to be kept away from what will be the final edge. In this laboratory we clean by vigorously rubbing this edge with a towel wet with acetone. Simply flooding the glass with an organic solvent is not adequate, and it is important to use a solvent which does not leave a film of impurity behind. Once this face has been cleaned, a series of knives can be prepared in succession from the strip of glass, but if there is any question about its having been contaminated it should be recleaned.

Only two tools are needed to break knives. The hardened wheel-cutters that glaziers use to score glass so that it can be broken properly are generally available in all hardware stores. One does need, in addition, a pair of glazier's pliers (Fig. 13(a)). These are designed so that they fit over fairly thick glass, and their straight broad lips can be lined up with the score mark made by the cutting wheel. Thus, they grip the glass and allow the break to be started in the proper place.

There are special tricks and precautions to be observed in

(c) The fractured face of a strip of glass which will be used to form the front face of glass knives. This is a very satisfactory piece of glass for such purposes since it is flat and without checkmarks. Two large chips near the upper part of the figure could be avoided easily. (d) Like (c) except that the fracture face of the glass is severely contoured, which would make it most unreliable as a front face for glass knives. Such a piece of glass should be rejected. (e) Like (c) and (d) except that "checkmarks" mar the face of this piece of glass. These coarse fractures almost certainly would intersect with any knife edge that might be broken. A few checkmarks in a piece of glass are acceptable if they remain clear of the knife edge.

breaking glass knives. Perhaps the most satisfactory procedure is to begin with a squared end. A score mark is made starting just behind the edge which is to be the front face of the knife, and just a little below the edge at right angles to this as illustrated in Fig. 13(b) and (c). Thus, as the knife is broken, the crack is expected to follow the score mark to its end, and then swing freely to intersect the edge that is to be the front face. This last and critical part of the fracture is not seriously constrained from above by a large mass of glass. Quite uniform, reasonably flat edges can be obtained as in Figs. 11(a) and 14(a).

Proper scoring of the glass is a minor art. Just enough pressure should be applied to the cutting tool to make the score line. The cutter must be held vertical to the glass. Scoring is done with a rather slow motion, no faster than one might draw a line with a pen and ruler. Successful scoring produces a slight sizzling noise as the glass is flaked. If this slight noise is not heard, scoring is not going well. The novice should practice scoring on scrap pieces of glass. One must take care to use just enough pressure to score. Greater pressure accomplishes nothing as far as the glass is concerned, and it is apt to flatten the steel edge of the cutting tool. One then can do nothing but throw the tool away. Even new cutting tools sometimes are too dull to cut properly. A good tool, however, will last for months if it is treated carefully.

To break the glass, what is to be removed is gripped with the glazier's pliers near the back edge of the glass strip (Fig. 14(a)). The pliers should be ¼ or ½ in. away from the back

---

FIG. 13. (a) The tools and glass necessary for making glass knives. In the middle is the scoring tool with a hardened cutting wheel, and on the right, a pair of standard glazier's pliers. The strip of glass has been scored at 45° as though to start making a knife. (b) A score line has been drawn almost to where the final knife edge will be. After breaking (Fig. 14a), the uppermost piece of glass will be discarded and another score line drawn where this glass is marked with wax crayon. A second break here will produce the parallelogram of glass which is the knife. (c) An enlarged detail of the score mark near the corner of the glass which will become the knife edge. Note that the score mark does not actually intersect the presumptive front face of the knife, and that its projection (arrow) intersects the front face 1–2 mm below the squared-off top. By placing the score mark in this position, the fracture which actually produces the knife edge breaks with a minimum of distorting stress from internal strains.

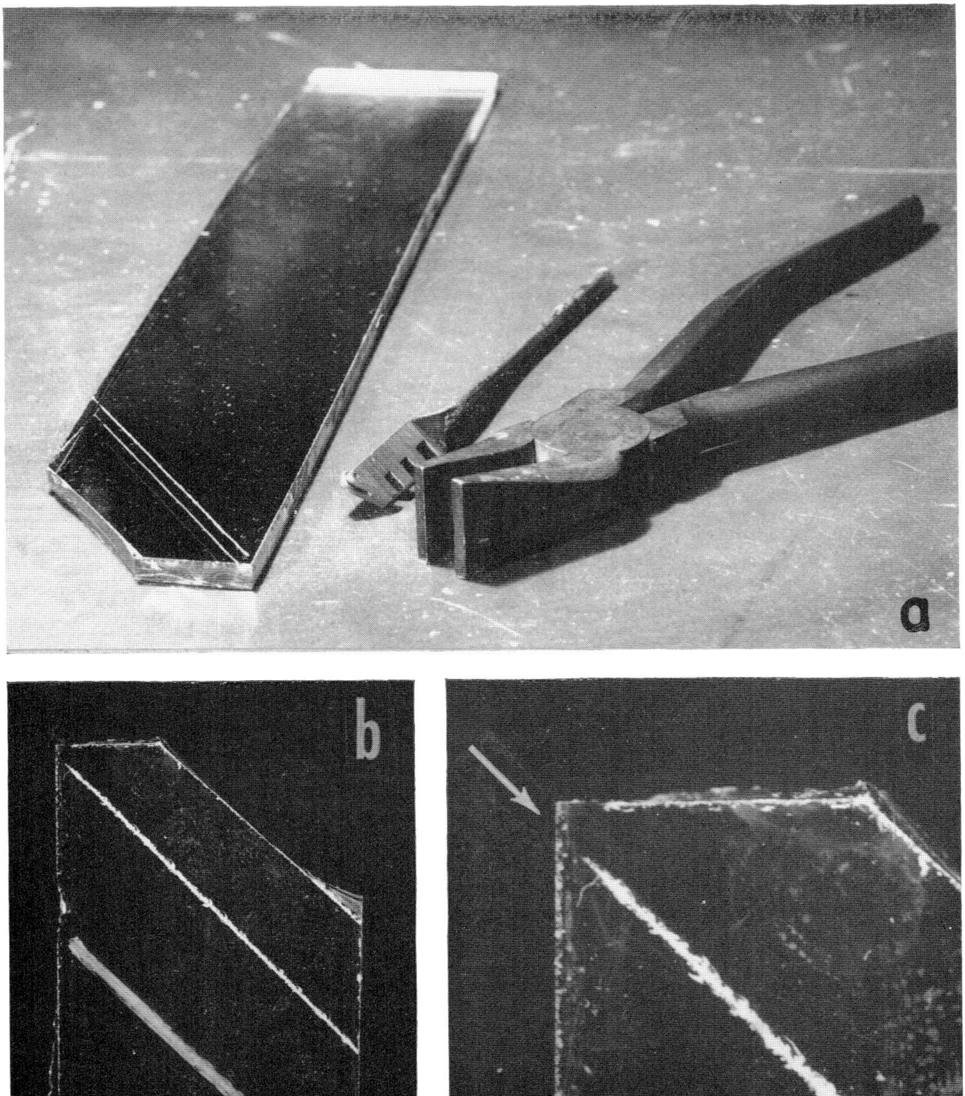
a
b
c

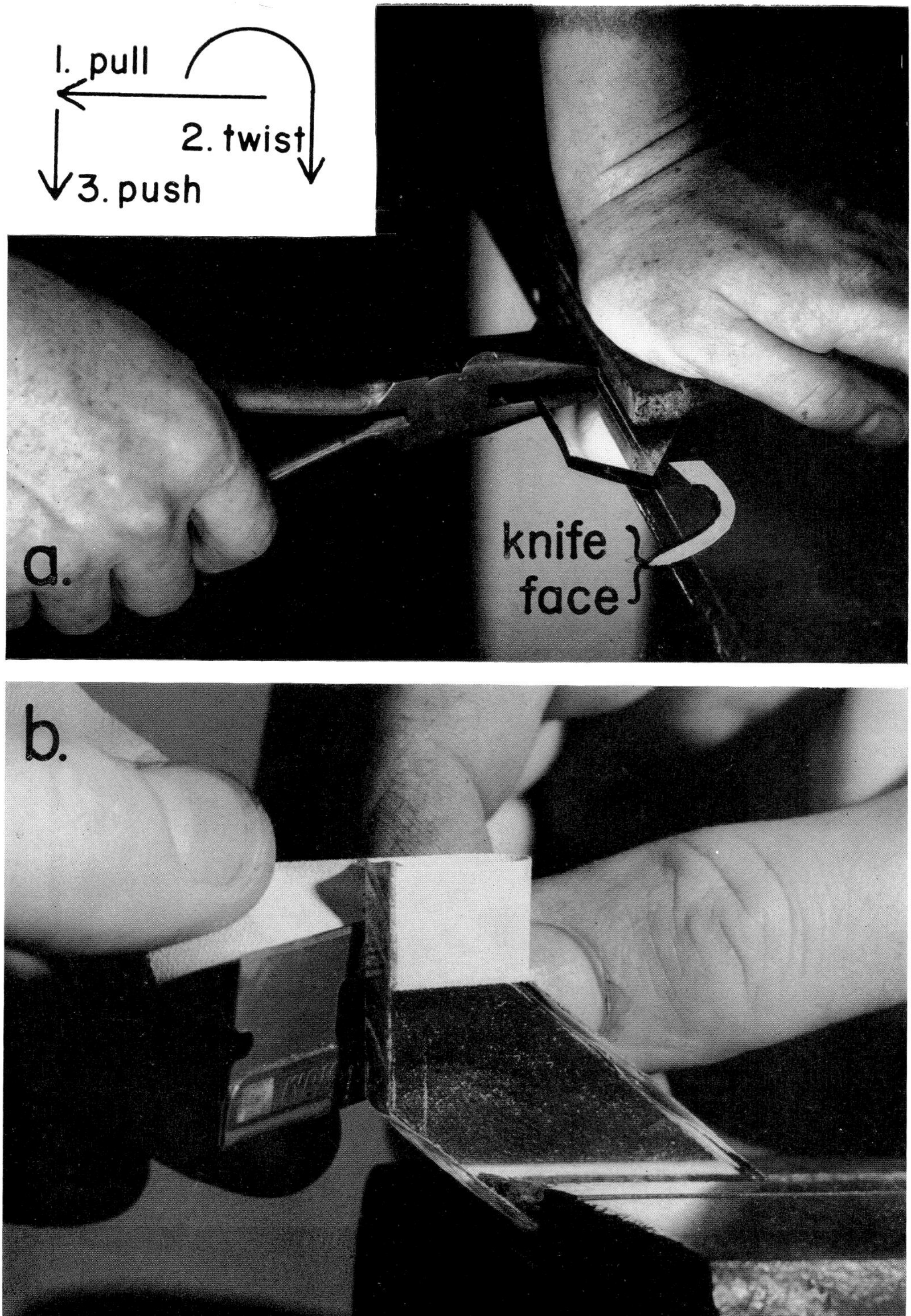
1. pull
2. twist
3. push
a.
knife
face
b.

edge. The upper lip of the pliers is placed directly along the score mark. The part gripped by the pliers overhangs a table edge. The strip of glass is held fast against the table top with a block of wood, but the block is kept clear of the corner which will produce the knife edge. Straight and flat edges (like that of Fig. 12(a) rather than 12(b)) will be produced most reliably if tension is established in the glass before it is cracked. Thus, in starting the break, one should first pull on the pliers horizontally, and then push down while maintaining that tension. At least 80% of the knives so broken should be quite adequate. If the novice has a much greater mortality it is probably because he has overlooked some detail, although very occasionally the glass itself may be at fault.

We have had pieces of glass from time to time which would not break properly along score lines, perhaps because they were poorly annealed. One generally has warning of impending trouble in that the cutting tool does not behave properly in making a good score line. Such glass usually breaks sufficiently erratically to produce steeply slanting knife edges. We have the impression that even when an edge of sorts can be obtained from such glass, it does not section well. Thus, with experience, "bad" glass is not too difficult to recognize and reject. "Good" glass scores and breaks cleanly and easily without great force, almost as easily as breaking a cracker.

FIG. 14. (a) Breaking the upper knife surface. A piece of glass is scored as in Fig. 13, and is laid on a table so that the score mark lies exactly over the edge. The glass is held down with a piece of wood, using the left hand. The glazier's pliers grip the dangling piece of glass near what is to be the heel of the knife. The upper lip of the pliers is laid exactly along the score mark, then the break is made. Knife edges will be more standardized and flat if a certain pattern of tension is set up as the break is made. This is diagrammed in the insert. One should first pull horizontally on the pliers; then the wrist should be twisted (supinated) so that tension is directed towards what will be the knife edge. Finally one pushes down, making the break. (b) A knife is held in the jaws of a vise while a trough is constructed of adhesive tape. The front face of the knife is mounted vertically and the adhesive tape wrapped horizontally. Excess tape is trimmed with a razor, taking care not to contaminate the front face of the knife. The trough will be made watertight by painting the heel region with beeswax. It usually is necessary to paint the inside of the trough with hot wax in order to avoid contamination of the knife edge with substances derived from the adhesive.

The method of breaking glass knives described above has been in use in this author's laboratory for about a year, and has consistently given reliable results. However, it does involve squaring the end of the glass strip before making each knife which is wasteful and often is unnecessary. Sometimes successive glass knives can be broken from a strip so that the break which creates the bottom edge of one knife is satisfactory as the top edge of the next. Thus, it is not strictly necessary to start with a right angle and break into it. However, then there is apt to be a high, and more or less objectionable spur left at the edge of the knife away from the scored side, and the whole edge may be tilted excessively (Fig. 12(b)). These undesirable features can be minimized by a suitable tension set up in the glass by wrist and arm action as the break is made. The diagram accompanying Fig. 14(a) shows the desirable pattern. In getting set to crack the knife, first establish a pull on the pliers in a horizontal direction. Then supinate the wrist strongly so that a torque is applied. Finally, bear down so that the glass is cracked across the table edge. Empirically, there is no doubt but that this combination of horizontal tension and torque influences the break advantageously so that flatter knife edges are obtained with a minimal spur. When one masters this trick, successive good knives can be broken from a strip of glass without squaring the end of the strip between most breaks.

Weiner (1959) has proposed a very simple method of breaking glass uniformly which seems most promising. He fastens a narrow (1/16–1/8 in.) strip of adhesive tape over the middle of the *lower* lip of his pliers. To break glass this is oriented just below the score line, with the lip about ½ in. in from the edge of the glass. (The tape is parallel to the score line in the glass, the lips of the pliers at right angles to it.) When the pliers are tightly squeezed together, the tape serves as a pressure point or fulcrum, and the crack starts and develops smoothly from here. Long strips of glass can be broken beautifully by starting the crack in this way, and then pulling strongly to extend the crack.

In some laboratories heat is used to break glass knives. The end of a heavy Pyrex rod is heated to its melting point with a compressed air or oxygen torch, and then pressed against the score line near the back end. This writer has found the method less reliable and less controllable than mechanical breaking, but he cannot claim much practice.

5.7. Angle of Knife

Knife angles of approximately 45° are commonly accepted as desirable for most purposes. This refers to the angle of scoring, and the major part of the break. However, as the break approaches the front face of the knife, it almost invariably is deflected so that it intersects the front face at a more obtuse angle. The actual angle at the edge cannot be known accurately, and no doubt varies considerably. Fortunately this is not critical.

Until recently it generally was assumed that difficult blocks of hard material had to be cut with knives having greater support than usual. Therefore, knives were made with more obtuse angles, 55° and 60° even being recommended. Of late, however, there has been a trend in the other direction, and it seems as though "support" may not be the most important factor. Thus, material embedded in hard Araldite or Vestopal W sections better when the knife angle is as acute as practical. Recent personal experience suggests that when there is difficulty with methacrylate blocks it is sometimes solved by using a more acute knife edge. The novice, unless he follows guide lines when he scores his glass, may gradually discover himself making more and more obtuse knives, which will add to his difficulties.

5.8. Glass Knife Defects

Knife defects can be so serious that ultrathin sectioning is quite impossible, or they can be so circumscribed that a few inconsequential scratches in the sections are the only result.

Glass knife edges are by no means uniform from one side to the other. The act of fracturing the glass by beginning the break on one side, sets up asymmetries which are quite constant in their general characteristics. There is almost invariably a curving "fracture ridge" which intersects the knife edge quite close to the side where the original score line was made (see Fig. 15(a)). The region of intersection should be avoided, and probably also any edge that may be available between this point and the near corner of the knife.

The other end of the knife edge (near the side of the glass which was not scored) often has a large or small spur sticking up (Fig. 15(a)). There almost invariably are fine check marks intersecting the portion of the edge in the vicinity of this spur. One can see these clearly by looking at knife edges with a dissecting microscope while tilting the knife to reflect light from this area. They are not particularly visible as the knife is actually set up for use in the microtome, but they can be seen if the operator knows what to look for. Where there are only a few small marks of this sort, they are not serious, and two or three widely spaced scratches in the sections might be the only effect. But when these checks are large and close together that region of the knife edge should be avoided. Usually checks of this sort are confined to about one third of the knife edge, including the region of the spur. Occasionally the check marks may involve most of the knife edge, in which case the knife should be discarded before use (Fig. 15(b)).

In practice, then, just a little more than half of a glass knife edge ordinarily is useful for ultrathin sectioning, and the rest of the edge must be avoided (except that it can be used for facing blocks). The inadequate parts of the edge result from the tensions and forces set up in the glass as it breaks. These can only be modified in minor ways by the manner of breaking or the

---

FIG. 15. (a) A good glass knife edge viewed from above. The spur to the right is small leaving much horizontal edge. This glass was scored originally on the left side. The spur invariably forms on the side opposite the score mark. A curving fracture ridge almost invariably can be seen arching towards the scored side, and commonly intersects the edge near that corner. The region of intersection should be avoided when sectioning although it may be used for facing blocks. (b) A similar knife edge, but marred by conspicuous checkmarks which intersect the edge along most of its extent. Such a knife should be discarded. However, one expects to find tiny checks in the vicinity of the spur so that one third or even half of the knife edge on the side opposite to the score mark ordinarily is useless for ultrathin sectioning (see diagram), and can be used only for block facing. The zone where one can expect the best edge is indicated in the diagram.

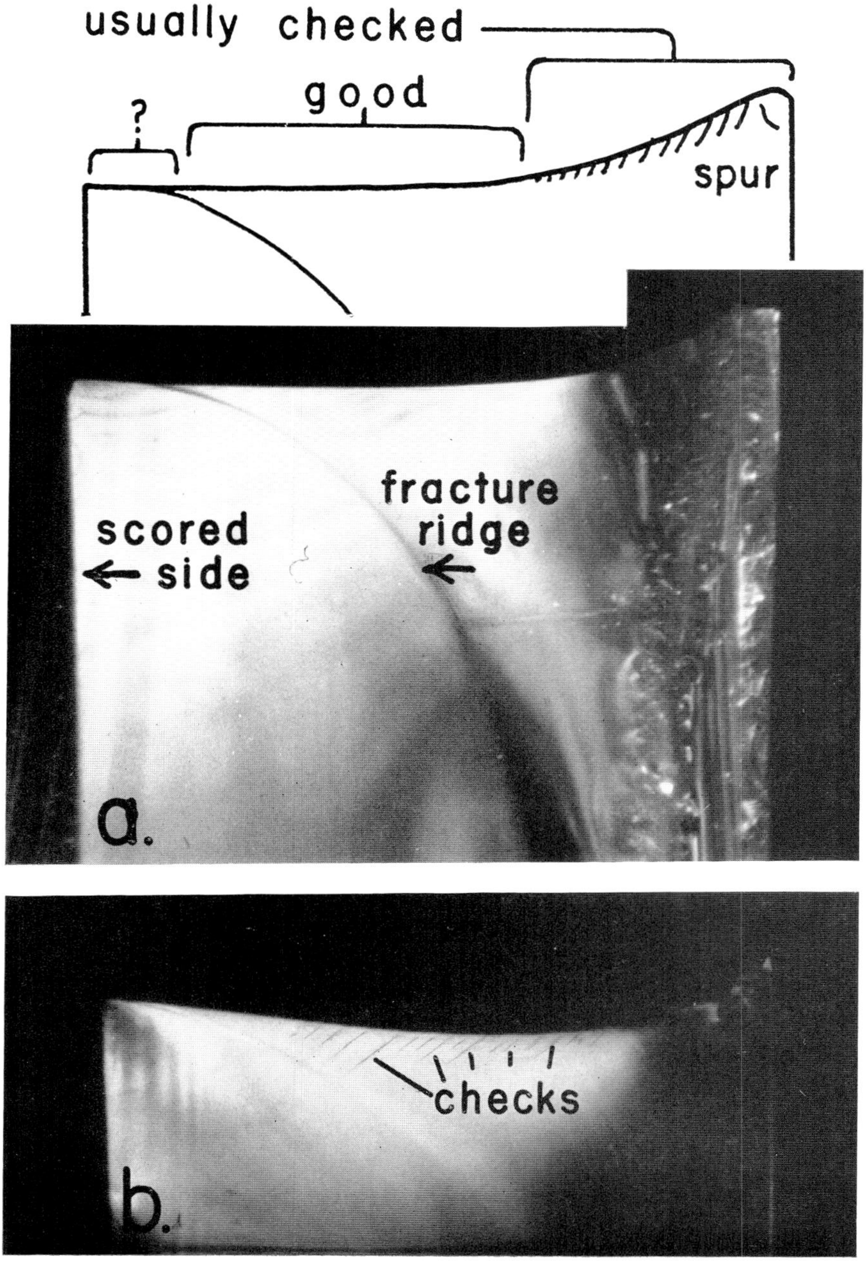
usually checked
good
?
spur
fracture
ridge
scored
side
a.
checks
b.

choice of glass, and all knives show the same general pattern, differing only in degree.

If knives are broken carefully, following the directions of the previous section, the edges should be reasonably horizontal, and spurs should be minimal (Fig. 12(a)). However, breaks will occur that produce tilted edges. Unless the tilt is extreme as in Fig. 12(b) it is not a serious defect. The principal difficulty in using a tilted edge is in keeping all portions of the knife edge wet during use. Actual cutting qualities do not seem to be affected, but the inconvenience is such that knives with excessive tilts should be rejected.

5.9. Diamond Knives

Diamond knives, as described by Fernández-Morán (1956) are now commercially available.* Knives with a fairly acute included angle of about 45° can be purchased which are suitable for most histological work. Knives with an obtuse angle of about 60° can also be obtained, but are recommended mainly for hard materials including the softer metals. But the latter may serve special biological purposes in dealing with mineralized tissues. Under the latter circumstances, the embedding material should also be hard, perhaps even pure methylmethacrylate.

An obvious advantage of diamond knives is that once the correct tilt has been obtained, the knife can be left undisturbed in its holder for repeated sectioning over long periods of time. Thus, it eliminates the uncertainty and inconvenience of positioning a knife every time sectioning is to begin.

Diamond knife edges are cleaned by rubbing them with a thin sliver of wood, wet with acetone or other organic solvent. The sliver is prepared by making a long diagonal cut through an applicator stick, a match, or a toothpick. The flat side of the thin wedge is then drawn gently along the edge so that it is ac-

* Diamond knives may be purchased from The Venezuelan Institute for Neurology and Brain Research, Caracas, from Ivan Sorvall Inc., Norwalk, Connecticut, and from E. Leitz, Inc., Wetzlar.

tually cut. This should be done while being observed with a binocular "dissecting" microscope.

Diamond knives are relatively hydrophobic so that when the trough is filled, the edge usually has to be wet with a sliver of wood. This means that a higher meniscus must be maintained with a diamond knife than when glass is used whenever everything else is equal. It seems likely that this would complicate the sectioning of Araldite and Vestopal W embedded material, for the tendency of the block face to become wet presents some difficulties with these substances. But this writer cannot speak of this from experience.

It is suggested by one manufacturer of diamond knives that blocks be somewhat harder than is customary for glass. This is accomplished by adding a little more ethyl- or methylmethacrylate to the embedding mixture, but the difference is not critical.

Although a diamond edge may retain its sharpness for a period of months if it is not chipped, eventually it must be resharpened. The distributors of diamond knives are prepared to do this.

One serious drawback to diamond knives is their fragility. The edge, being thin, is reasonably flexible, but it can be chipped through carelessness. It is most important to avoid rubbing it at right angles, for instance by jamming it into a block or chuck when positioning the knife holder, or by touching it with a pair of forceps or a needle while working in the trough area.

5.10. Trough Construction

After a satisfactory glass knife has been prepared, the next step is to construct a trough. When this is filled with a suitable fluid which wets to the knife edge (usually a dilute acetone solution), sections are floated as they are cut. (Diamond knives are permanently mounted in a holder which incorporates a trough.)

The simplest way to prepare a trough is to use an adhesive tape and wrap it from one side to the other as illustrated in Fig.

14(b). Excess tape overhanging the face of the knife is carefully cut away with a razor (Fig. 14(b)) taking great care to avoid contaminating the cutting edge or the front face of the knife.

In making the trough, the knife should be mounted for convenience in some sort of a vise. Its front face should be vertically oriented, then the top edge of the trough can be accurately aligned at right angles to this by holding and pulling the tape horizontally. A correctly designed trough then is made almost automatically.

It is a mistake to make too large a trough. It will become apparent later in this chapter that sections can be handled most easily if they are in a rather confined space. Therefore, rather narrow adhesive tape should be used so that it does not extend far along the back of the knife. If the knife angle is approximately 45°, the tape should not be more than ½ in. wide and preferably about ⅜ in.

Naturally adhesive tape cannot be made to fit snugly across the irregular back of the knife. Therefore, this region of the heel of the trough is filled with wax to make a seal. Filtered beeswax is very effective. This wax is tacky and adheres to both the glass and the adhesive quite securely. It has a low melting point, and a pot of it can be kept melted easily in any paraffin oven. It does not contaminate the acetone solutions used in the trough. A small camel's hair brush is kept in the pot, and the hot wax built up both inside and outside of the trough in the region of the heel to complete the seal.

The adhesives used on adhesive tapes are a potential source of danger as far as contaminating the knife edge is concerned. The adhesive should be suspect if the glass near the knife edge becomes hydrophobic as it is used. This can be avoided if the interior of the trough is painted with hot beeswax. Waterproof adhesive tapes can be found, however, which do not contaminate, and can be used without this precaution. Manufacturing companies have been "improving" and changing their adhesives in recent years so that a tape which is satisfactory at one time may not be when reordered.

Although the above method of making troughs seems quite reliable and easy, many laboratories prefer to use metal troughs made from sheet copper or aluminum. These are cut and bent appropriately and are fastened to the knife with molten beeswax or dental wax softened with a hot needle. One trough, of course, can be used over and over again if moderately heavy sheet metal is used. This seems a small advantage when compared with the difficulty of putting on the trough without damaging or contaminating the knife edge. Foil may be used and then discarded. Probably the chief justification for using metal is that when properly cleaned it will not contaminate the acetone solution. Cleaning, of course, means that grease films are eliminated, including those that might come from fingerprints.

A "trough" of sorts can be made most easily simply by drawing a line with a wax glass-marking pencil across the back of the knife. The line is parallel to the edge and about $\frac{1}{4}$ in. behind it. This will confine a drop or two of acetone solution to the edge region of the knife and the sections can be floated on this surface. Naturally, one cannot cut very many sections before the surface is covered, and there is no good way of picking them up except directly on a screen grid. Working in such close quarters, one is apt to damage the knife edge in doing this, and with so little fluid there are added dangers of contamination. Furthermore, the acetone evaporates rapidly. For these various reasons the method is not recommended for general use, but it is rapid and convenient when only a few sections are needed, perhaps for test purposes, and when one has no interest in conserving the knife edge.

On occasion one wishes to move sections about within the trough, this may be done with a sharp toothpick having a bent tip. The tip should be wet in advance in the acetone solution, and introduced under the surface of the trough fluid. The bent tip is then delicately brought up underneath the section to be moved until it touches. It is then possible to move the ribbon and, as importantly, disengage it afterwards.

5.11. Mounting the Knife

It might seem that mounting the knife in the microtome hardly deserves a section in this book, yet the proper knife tilt is a most important thing for successful sectioning and care must be taken in determining it. Also, insofar as possible, it is important to recognize the symptoms of improper tilt so that suitable corrections can be made while sectioning is in progress.

A slight tilt is necessary; about 3 to 5° seems best (Fig. 16).

Too much tilt encourages "chattering" (Fig. 17) which is the technical bugaboo of ultrathin microtomy, for the chatter is often on such a microscopic scale that it is not visible while sections are being cut. It is disappointing to discover when looking with the electron microscope that seemingly well cut sections are quite useless for this reason.

Too much tilt probably promotes chatter because the knife edge tends to dig into the block, and then to bounce back when elastic forces become great enough. One should recall how flexible thin glass is as we experience it in a fine capillary or in glass wool. It is a wonder indeed that the very fine edges of glass knives can have enough support to cut at all. They gain maximum support when the front face is as nearly vertical as possible, and the main mass of the glass can absorb the vertical thrust.

Although one can be deluded into thinking that he is sectioning well when the knife is tilted too much, no such illusion is possible when the knife is leaning slightly backwards. Then the block coming down misses the edge and rubs against the back of the knife. This compresses the block a little and the same thing happens again on the next cycle. Thus, no sections at all are cut for a number of cycles of the microtome, but then suddenly a thick section is cut when the block recoil catches up with the additive effects of compression, and the knife digs in. Thus, when this is happening, there is excessive skipping and occasional thick sections. Also, one can sometimes see the block rubbing on the glass by looking through the acetone and through the glass to the front face of the knife. Where the block makes contact with the glass the reflectivity of the glass surface is altered and a black spot may be visible.

How then is one to tell the correct angle? There is no simple answer to this except to set the front face as nearly vertical as one dares (Fig. 16). To do this it is absolutely essential that the operator bestir himself from his comfortable chair in front of the microtome and move around the side of the instrument, some distance away. With this perspective he can judge the verticality of the front face without much difficulty.

It is, of course, possible to rig jigs which will set the same gross angle of the knife face every time. This does not solve all problems, however. The block in the Porter–Blum microtome actually swings in an arc. If the knife edge happens to be mounted relatively high, a little more tilt is required to maintain the same tangential relationship than if the knife edge is low. Also, the front face of the knife is fractured glass and is never perfectly plain. The reader has already been warned not to use a glass face which is decidedly wavy in making a knife, but one cannot know exactly how the face may be oriented in the immediate vicinity of the knife edge. Thus, standardizing the gross angle can only give an approximation of the important clearance angle in the vicinity of the cutting edge. Once one understands the situation, it is preferable to err, if at all, by making the knife too nearly vertical. Then, as one starts to cut, the symptoms of backward tilt may appear. Remedial measures are easy. Tilting the knife 1 or 2° may be enough to make all the difference between no thin sections and perfect cutting. This angle actually is very critical. However, fortunately there is safe leeway of perhaps 5° beyond this point.

In mounting the knife one must observe yet another precaution. In the Porter–Blum microtome there is a vertical bearing along which the cantilever arm slides as sections are cut. This bearing is straight and vertical for only about ¾ in., and perhaps only the middle ½ in. of this should be used. Thus, the knife must be mounted with its edge in such a position that sections will be cut during this part of the cycle, and not when the specimen is either too high or too low (see Fig. 16). The novice is apt to have the knife sticking diagonally out and up too far. One might avoid this by standardizing knife sizes and the manner

of mounting, but it seems easier just to check the edge height each time the knife is mounted. This allows flexibility of operation, for occasionally one wishes to tilt the block decidedly up or down which would complicate any attempt at standardization.

5.12. Preferred Orientation of Block in Microtome

In mounting the block in the microtome, usually the primary consideration should be to minimize the forces which tend to deform the knife edge. Therefore, in general, the short axis of the block face should be parallel to the knife edge. The narrower the block, the less chance there will be for "chatter" and "digging"!

Dense connective tissue with its quantities of tough collagenous fibers is particularly difficult to cut smoothly. If this involves only part of a block it should be placed on the trailing side. Thus, to cut skin, epidermis should reach the knife edge first, followed by the dermis. In cutting a portion of a large artery, endothelium should be first and the adventitia last, etc. In this way coarse chatter effects and tears are at least confined to the connective tissue zone, and the rest of the tissue spared.

On occasion one wishes to make transverse sections of a thin sheet of tissue so that the rectangle of the block face becomes very disproportionate, with one very narrow axis, and the other as long as possible. It may be impractical to mount the long axis

---

FIG. 16. The region of the knife and block holder of a Porter–Blum microtome. A glass knife with a trough made of adhesive tape and sealed with wax is in its holder. Note that the front face of the knife is nearly, but not quite, vertical. Also, the knife is snugly down in the holder for maximum support and to prevent its edge from being at too high a level. The block must pass the knife edge during the middle portion of its downstroke.

This figure indicates how specimens may be removed from the trough using a glass spatula as described in Chapter 5. If the glass previously is wet with the same acetone solution as is in the trough, surface tension will pull sections up on the glass as acetone rapidly evaporates from the thin film on the spatula.

This microtome has been modified slightly in that a V-groove was cut in the fly wheel, and a leather belt fastened to the table at one end dangles over it. This serves as a brake as described in Chapter 5. A massive chuck also has been provided which adds inertia to the system. The microtome is being used with a binocular dissecting microscope mounted on an extension arm and having a very long focal length objective, visible in the upper left of the figure.

vertically, and it would be far too wide to mount it parallel to the knife edge. Such very narrow blocks can be cut by mounting them with the long axis somewhat oblique to the knife edge. As the knife sections, then, only a small part of its edge is being used at any given moment. This technique combines nicely with the problems discussed in the previous paragraph. This writer, for instance, has successfully sectioned long sheets of the anterior cornea by mounting them diagonally, with the epithelium facing the knife edge, and the lamina propria trailing.

Ordinarily the side faces of a block will be nearly parallel and should be oriented vertically. If this is not the case, an overhanging corner of one section may not be pushed cleanly away from the knife edge by the succeeding section, and sections will not spread smoothly over the fluid surface of the trough. They may be severely folded at one end and tend to form a circle rather than a straight ribbon. This can be avoided by trimming the block so that the lower edge is slightly longer than the upper edge and the lower corners stick out laterally farther than the upper corners.

5.13. Section Flotation

If methacrylate sections are given a chance, they will stick with great tenacity to dry clean glass. Thus, one cannot collect them on a dry knife surface. The introduction of a trough and flotation fluid by Gettner and Hillier (1950) was a most useful development.

A desirable fluid to use for section flotation would seem to be one which wets glass fairly vigorously, yet does not wet clean methacrylate. Dilute acetone solutions answer these requirements well and are generally used. Different laboratories, however, use varying concentrations of acetone from perhaps 10–50%. In our own laboratory we prefer 40%. This fairly strong acetone solution has secondary advantages in making it easy to remove sections from the trough in order to transfer them to other dishes (discussed in Chapter 5.21).

In general, one uses a nearly flat meniscus at the knife edge for sectioning, although this depends somewhat on the angle of the knife. Thus, quite a highly convex mensicus may be necessary when the knife has an obtuse angle, while a somewhat concave meniscus will work best with an acute angle.

Considerable latitude is permissible in the shape of the meniscus. If it gets too high, however, the block face will be wet and fluid will be carried over the edge of the knife and down its front face. This, of course, is apt to contaminate the edge. In any event one has to wait for the face to dry, and it is at the very least a considerable nuisance. On the other hand, if the meniscus is too low the sections will not glide away from the knife edge in an unimpeded fashion. There will be some back pressure and a tendency for sections to rumple and not flatten out properly. In extreme cases, of course, they will pile up on each other.

In starting one's sectioning it is desirable to have the meniscus low rather than high. The block then can be faced without being wet and the edge contaminated, and a ribbon of sorts can be started. It is then easy to add a drop or two of acetone solution and get successive sections which glide smoothly away from the edge.

Occasionally, as one attempts to section, the block seems particularly prone to being wet. Sometimes such blocks can be successfully started by lowering the meniscus to the point where there is incipient drying of the edge. Then a few rumpled sections may be cut which are at least dry. More acetone is then added judiciously to raise the meniscus a bit until smooth expanded sections follow. These difficulties occur mainly when the knife is broken in such a way that the actual edge angle is decidedly obtuse, and therefore requires a relatively high meniscus.

Acetone is a possible source of several contaminants and only reagent grade acetone should be used, diluted with demineralized or distilled water. Otherwise various oily films may appear which not only are likely to contaminate the knife edge, but may leave very messy deposits on the sections so that they become almost too dirty to be used finally. Also, some tap waters may precipitate mineral salts which will dirty the sections.

When the epoxy resin Araldite or the polyester Vestopal W

is used instead of methacrylate as the embedding medium, pure water or very dilute acetone (less than 10%) should be used in the trough. Stronger acetone solutions tend to wet the blocks, and the sections themselves then nosedive under the surface as they are cut.

5.14. Flattening Sections

The acetone solutions upon which methacrylate sections are collected produce a considerable amount of flattening. It has been noted, though, by Morgan *et al.* (1956) that sections may remain compressed to some degree so that structures which should have circular profiles are found to be elliptical, with the short axes parallel to the direction of sectioning. To a large extent this can be eliminated by softening the methacrylate somewhat with the vapor fumes of a volatile solvent, whereupon surface tension will produce fuller expansion. Thus, a fine camel's hair brush or a wisp of cotton on a toothpick dampened with xylene can be brought judiciously near sections as they float in the trough upon the acetone (Satir and Peachey, 1958).

In this author's laboratory solvent vapors have not generally been used to flatten sections because it has been common practice to transfer sections from the trough to containers filled with 50% alcohol. This also very nicely flattens and extends the sections so that nuclei and other globular objects look circular when seen in the electron microscope. (The manner in which this is done is described in Chapter 5.21. The technique was originally employed so that sections could be selected carefully, and grouped so that a number of good sections could be mounted on a screen. Also this affords an excellent opportunity for "electron staining.")

5.15. Sectioning

Many people have difficulty in mastering the art of ultrathin sectioning, yet others learn this without serious trouble. It is not so much an art as it is a skill requiring much finesse and attention to detail. One must adopt the attitude that one is going to do the utterly impossible in cutting sections only a few millimicrons thick. Yet the beginner must be buoyed up by the knowledge that thousands of excellent electron micrographs have demonstrated that it can be done. It cannot be done, however, by working as one does in conventional microtomy. It is almost a disadvantage for the novice to know conventional methods for he has bad habits which must be eliminated, as well as an attitude of mind about the work that absolutely must be changed.

The unaided human eye is not enough for ultrathin sectioning. From the time the knife is mounted in the microtome until after the sections are finally mounted on support grids all operations must be performed using a binocular "dissecting" microscope. The person without experience using this instrument may have some difficulty at first with his manipulations, but he should rest assured that with practice he will learn where his fingers are.

As a preliminary to sectioning, the microtome is cycled and stopped so that the block is in approximately the same horizontal plane as the knife edge. The knife is brought fairly close to the block face so that both are visible in the same field of the binocular viewing microscope. The binocular microscope is adjusted so that the point where the block and knife edge will intersect is a bit above the center of the field, and is in focus. Thus, there will be a good view of the most forward part of the trough.

At this stage of the game, the knife will probably need a lateral adjustment so that sections will be cut from the portion of the edge that one desires. Since the first cuts are apt to be thick sections to face the block, a waste portion of the knife may be used, far to one side or the other, leaving the middle section for quality sectioning later. If a good portion of the knife edge is used for moderate facing of the block, and this is done with delicacy and without taking too large a number of sections, it may be possible to continue with ultrathin sectioning without moving the knife.

Coarse and fine adjustments allow the knife to be brought up to the block face. The perspective with which one views the block and knife is such that it is not always easy to see just how close the knife is to the block face. Often, however, the light source can be adjusted so that the knife edge will cast a thin dark shadow on the block face if the latter is dropped a little below the edge of the knife. The narrowing shadow observed as one manipulates the controls indicates the closing gap between the knife and block. Under no circumstances should the knife actually be allowed to touch the block, for most of the time this will so contaminate that portion of the edge that it must be abandoned.

When the block and knife are as close together as one dares place them further advance is by cycling the microtome while still advancing the knife by small adjustments. Eventually the microtome will start cutting "thick" sections, although the advance should be so cautious that these "facing" sections should be only about a micron thick. If thicker than this the knife may dig and gouge, or chip out portions of the block.

The most convenient time to fill the trough is just after the knife has made contact with the block. The acetone solution is added with a pipette until the edge is wet. So long as thick facing sections are to be cut, it is best to keep the meniscus low to minimize the danger of wetting the block face.

When final facing is completed and one is ready to begin thin sectioning it is usually desirable to get rid of the thick sections floating around in the trough, and to clean the surface and adjust the level of the acetone in preparation for the ultrathin sections. This is done with a glass spatula previously wet with the acetone solution, in a manner described in detail in Chapter 5.21.

If we assume that there has been an interruption to clean the acetone surface, thin sectioning is then begun by first taking one or two slightly thick sections by delicate manipulation of the fine adjustment of the knife holder. Then one continues on, simply cycling the microtome. Success will depend in part upon factors already considered: (1) a clean knife, (2) a proper tilt, (3) a meniscus at the proper level, and (4) a sufficiently small block, one half the size that the novice hopes to cut. Success also depends upon three other factors: (1) the proper (slower than you think) cutting speed, (2) a steady cycling rhythm, (3) a light touch.

The cutting speed is quite important. Fortunately there is a simple way of judging this. If you can follow the movement of the block with your eyes as you look through the binocular viewing microscope you are cutting slowly enough. There is a strong unconscious tendency to speed up the cutting when apparently sections are being cut successfully, but excessive cutting speeds lead to the fine-order chatter discussed later. Thus, even though you may get sections with fast cutting, these are apt to be worthless. It is possible, but unlikely, to cut too slowly.

A steady rhythm is essential for successful ultrathin sectioning. The microtome can, and should, be cycled fairly rapidly except during the actual cutting phase, for a short period favors reproducibility, no doubt partly by minimizing thermal fluctuations. A comfortable rhythm that can be maintained is perfected only with practice. Until this skill is developed, sections will be more variable than otherwise might be the case.

In general, then, the cycling motion is divided into two phases. The first is the fairly rapid phase which begins after a section has been cut and the block has disappeared from view behind the knife, and which ends perhaps with a momentary pause as the block is poised above the knife, ready to come down across the edge. The second phase is the slow vertical descent of the block across the knife edge. It is, of course, important that this last be as steady and as controlled a motion as is possible. This author thinks that the operator's hand and arm should be either pushing or pulling in the horizontal plane during the second phase of the cycle, and therefore recommended in Chapter 5.3 that the microtome handle be rotated on its shaft to establish this correspondence.

The importance of maintaining a steady rhythm is at once apparent after there has been an interruption for one reason or another. Thermal changes will have occurred during the interruption and the first cycle may produce a thick section, or conversely, nothing, and the knife may have to be moved into the block face with delicate adjustments. The net effect of this will also be to produce a thick section. These thick sections, no doubt, set up undue compression effects within the block, so that as one begins a sectioning rhythm the first sections are apt to be of rather varying thickness, and there may be considerable skipping. *It is only after perhaps a dozen or more cycles that the microtome and the block will settle down so that reasonably uniform sections can be cut.* The experienced operator, thus, hates an interruption when he is cutting well. One sort of interruption is inevitable when enough sections have been cut so that the ribbon bumps into the trough edge, or into rafts of other sections and starts showing signs of serious back pressures which prevent the proper expansion of new sections. Then an interruption must be accepted to remove the sections from the trough. More annoying is a telephone call which had better be ignored while good sectioning continues. Still more annoying is someone in the laboratory who wishes to talk and divert attention. The operator will do well to do no more than grunt replies without looking up.

A "light touch" is very important in ultramicrotomy. In thinking about this one should bear in mind that the sections he hopes to obtain will be at most only a few hundred Ångström units thick. One naturally does not want much thickness variation within a single section. The machinery of the microtome cannot be allowed to deviate more than perhaps a hundred Ångström units during the cycling or the actual sectioning. We are approaching the dimensions of modest sized molecules, and we are faced with the problem of keeping within these tolerances in quite a bulk of microtome. Even massive castings can be expected to bend this much when moderate pressures are applied. Quite apart from theory, it is empirical experience that the operator can distort the microtome enough to ruin thin sectioning simply by keeping a tight grip on the microtome handle, and cranking it around as he might be used to doing in conventional microtomy. A truly delicate touch is indicated, and only fingertips should touch the handle lightly during the cutting portion of the cycle. Indeed, this author has developed the trick of using only the ball of the thumb during the cutting phase of the cycle, and pushing against a small back pressure created by a drag which has been described earlier. This is his only contact with the microtome during this critical part of the cycle. It is easy to forget to maintain a light grip and regress to longer-established patterns.

Physical deformation of the microtome, of course, may be brought about by extraneous vibrations, some of which have already been considered. The operator, though, may be guilty. He must not have his elbow on the microtome table; nor should his foot be hooked over a table rung. The operator, in fact, should have no contact with the microtome table during sectioning other than his fingertips on the handle of the microtome itself. Under these circumstances, the operator is forced to use his shoulder and elbow joints rather than his wrist in the cycling of the microtome. In spite of this being rather tiring, it is highly desirable; for an unsupported wrist does not seem capable of producing sufficiently delicate movements at the slow tempo required for sectioning. On the other hand, the shoulder does this admirably.

Good sections, of course, must be sufficiently thin, and this can be ascertained by observing their interference colors as indicated in Chapter 5.17. Very importantly, too, sections must be free of the fine-order chatter* (Fig. 17) that has been mentioned already in relation to the embedding medium, knife tilt, and block size. It is not always possible to recognize this chatter in sections as they are cut. It seems safe to say, however, that "glassy" sections will be free of chatter. Sections with a "frosted" appearance are suspect. Yet the appearance of good sections depends somewhat upon the nature of the embedded material, and under some circumstances one can have a frosted appearance and yet have unchattered sections. The frosted look, no doubt, represents light being reflected by interfaces within the section. These can either be the cracks and disturbances created by chatter, or they can legitimately be reflections from refractile material within the specimen.

* The exact cause of various sorts of "chatter" has been discussed, particularly by Morgan *et al.* (1956), and also by Williams and Kallman (1955) and Gettner and Ornstein (1956). It is apparent that at least three different phenomena may be distinguished, and quite possibly not all of them are truly vibrational in origin. For the finest order disturbance, Morgan *et al.* proposed the term "periodic distortion," implying alternating zones of compression and zones of good cutting.

a.

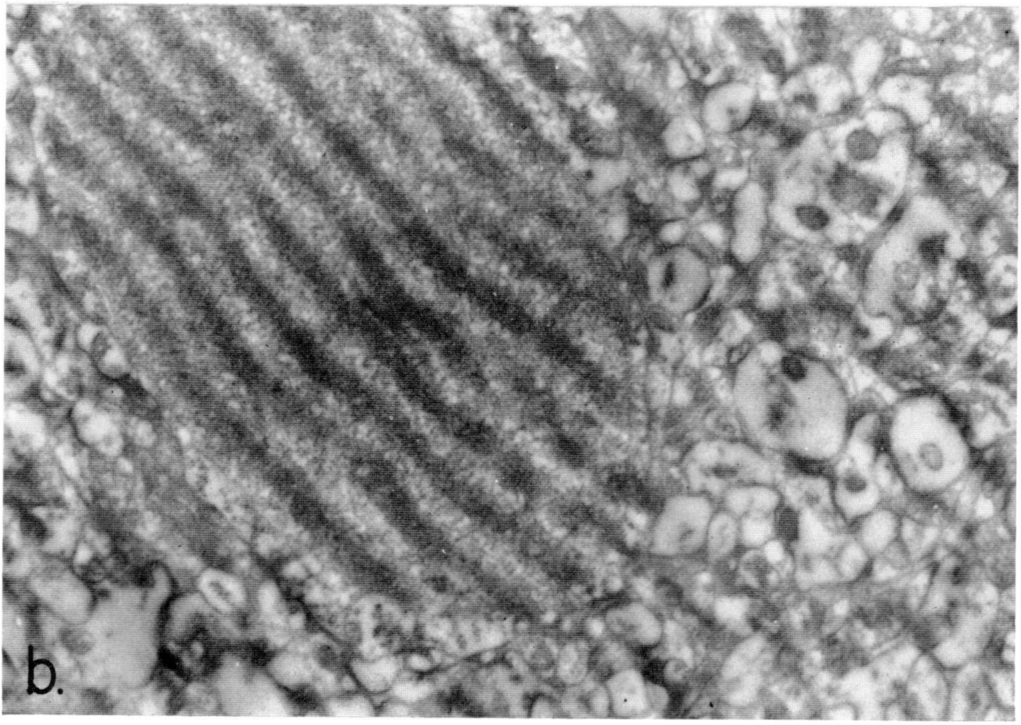
b.

If one suspects chatter, there are two things to consider that can be corrected without seriously interrupting sectioning. An oversized block is by far the commonest reason for chatter, and an operator with any experience at all may realize that consciously or unconsciously he has overstepped the limits. It is, of course, quite possible to trim, and particularly to narrow, a block without removing it from the microtome holder. The other thing to consider is the knife tilt, which often needs some adjustment to get good sectioning underway.

The beginner is apt to have a little difficulty establishing a smooth and slow cutting stroke. At first he is not sufficiently relaxed and his tight muscles make spastic movements. This may produce very coarse chatter in the sections so that their interference colors show obvious banded patterns. This is not too serious for there may be reasonable numbers of thin zones mixed with thick ones, and these will be satisfactorily sectioned.

Scratches, of course, may show up (Fig. 17(a)). A few scratches are not of importance for one ultimately can work between them satisfactorily. Many scratches ordinarily mean a contaminated knife edge which will have to be abandoned. If one sees a few scratches at first which quickly become numerous, the operator should suspect that dirt is present either in or around the specimen. The wax coat which may be in use to support the block (Chapter 4.10) is easily contaminated. It fortunately can be removed and replaced if necessary.

5.16. Contamination—Principal Cause of Sectioning Difficulty!

Contamination of the knife edge is by far the most common and insidious deterrent to success. The inexperienced investigator is not apt to recognize the warning signals, or appreciate all of the traps.

The precautions necessary to avoid contamination while a

FIG. 17. (a) A low magnification micrograph of a small arteriole, the sectioning of which was ruined by "chatter." Vertical knife scratches are also evident. (b) Tissue of the cerebral cortex at fairly high magnification. Chatter marks show prominently in the oligodendrocyte, but are much less evident in the adjacent neuropile, although still present.

knife is being made have already been considered. The walls of the trough also have been indicated as a potential source of trouble. In moving sections about in the trough, there is real danger of introducing oily contamination from dirty tools. One has to be careful of the acetone (reagent grade) solution itself and the containers in which it is stored. In general, once an oil film has been deposited on the edge of the knife it becomes worthless, and one can only start over.

Also, methacrylate itself can be a "contaminant." If the knife edge is allowed to become dry and then thin sections are cut, a section or part of a section may get stuck to the glass and adhere so vigorously that it cannot be floated free. Subsequently the acetone will not properly wet that portion of the edge. In these circumstances, it is usually possible to move the knife so that a new portion of the edge is used.

Simply drying an edge that has once been wet with acetone is apt to destroy its usefulness. Such dried edges usually become hydrophobic indicating that the acetone left an organic deposit behind as it retreated. Thus it is essential that once an acetone meniscus is brought to the knife edge, it be maintained there. If the operator leaves a wet knife to answer a telephone, it may be too late when he gets back. If a wet knife must be left briefly, it is wise to build up a high meniscus with surplus fluid in the trough that later can be removed as necessary.

Considerable areas of hydrophobic glass or oily films may be seen easily if the investigator is on the lookout. Often, however, the knife edge can be so locally contaminated that nothing can be seen with the usual low-power viewing microscope. This creates a problem for the beginner, but usually not for the experienced investigator. The latter will be confident that everything but the knife edge must be in order, and that he should be getting good sections if it were not for contamination. Instead, he may be getting "sections" that are crumpled, scraped, scratched, split, or shredded. Perhaps these difficulties started after an interruption, so he knows the knife was good to begin with. In any event, he expects a new knife to take care of the situation. The novice can have no such confidence for he is uncertain about so many other aspects of his technique, but he should always think of contamination first, and be prepared to break lots of glass!

It seems fair to say that a knife edge deteriorates almost entirely by contamination. If no grit is present to actually chip the edge, a knife can cut soft tissue almost indefinitely. On one occasion, as an experiment using a microtome with a motor drive, the present writer cut sections continuously for over an hour without any discernible deterioration, although several thousand slices had been made. Thus, when sectioning is going well, there is no advantage in shifting to a new portion of the edge, but when sectioning is going poorly it may be desirable to do this without delay.

5.17. Estimation of Section Thickness

In the early days of ultrathin sectioning for electron microscopy it was a time consuming process, taxing one's patience, to scan screen after screen with the electron microscope before finding sections thin enough to be useful. Our efficiency was vastly increased when it was realized that interference colors could be used in estimating section thickness, and so one could have the foreknowledge that suitably thin sections were being cut and mounted. This discovery was first formalized by Porter and Blum (1953).

Peachey (1958) has presented the most detailed correlation of section thickness and interference colors. His data are summarized in the following table.

Color	*Thickness* (Å)
Gray	600
Silver	600–900
Gold	900–1500
Purple	1500–1900
Blue	1900–2400
Green	2400–2800
Yellow	2800–3200

These values are higher than previous estimations; however, they are in approximate agreement with the measurements of Bachmann and Sitte (1958). Huxley (1957) working with muscle filaments, dealt with a system having an internal standard of measurement. He was able to obtain a group of "dark gray" sections approximately 150 Å thick, and another group 250 Å.

One sees interference colors when the surfaces of the sections are examined with a tilted microscope, and light is reflected from them. Thus, the binocular microscope used with a Porter–Blum microtome should be arranged with a decided tilt (Fig. 15). This is not only a satisfactory arrangement for visualizing the interference colors of the sections as they are cut, but it is also convenient for viewing the trough and the block, and for manipulations in that area. The light source should be adjustable so that it easily can be arranged to reflect light from the meniscus of the fluid in the trough. As the meniscus is changed it will be necessary to move the light to maintain the proper reflection. If the knife edge has a peculiar shape, perhaps a steep slope or a high spur on one side, the meniscus may have such a complex geometry that viewing the sections in reflected light becomes difficult or impossible. For this reason, knives with horizontal edges and small spurs are desirable.

The best light source for viewing sections by reflected light is a fluorescent lamp. Since the meniscus is not usually flat, a broad source of light helps. One also wishes to avoid the thermal effects of an incandescent light source close to the block and knife. The new miniature fluorescent lights on a movable arm are probably best. The entire fixture can be fitted in between the binocular microscope and the housing of the microtome, and casts a diffuse light over the whole trough area.

There is no simple answer as to what is desirable section thickness. For survey work or histological work when relatively low magnifications and resolutions are to be used, one wants much thicker sections than for high magnification, high resolution work. The very thin section does not have adequate contrast for survey work (Fig. 27(b)), as discussed in detail in Chapter 8.2. Therefore, in most situations, one is happy to have a series of sections showing a range of thickness. We have known for some years that gray, silver, and pale gold sections spanned this useful range. Deep gold colors are usually too thick to serve any useful purpose, and purple sections certainly are.

Commonly, when one is cutting as large a block as he can get away with, and when the Porter–Blum microtome is set to advance two units with each cycle, there is considerable variation in section thickness, so that one has gray, silver, and light gold sections produced together. When the microtome is set at one unit of advance, all of the sections may be gray and pale silver. One often can pick up groups of silver sections on one screen, pale gold ones on another, and a series that are mostly gray on yet another. If one really wants to obtain sections of great uniformity, for instance for serial section work, the block face must be very small and all other conditions of sectioning be favorable.

The beginner probably should concentrate on pale gold sections for there is no point in his trying to do high resolution work until he has mastered the variety of skills necessary to achieve this. It is fairly easy to obtain a good micrograph at 50,000 diameters, but it is quite another matter to achieve real resolution at 100,000 diameters or more when the investigator is approaching a whole host of technical limitations which conspire to defeat him.

5.18. Sectioning Araldite and Vestopal W Embedded Material

Araldite and Vestopal W have quite similar wetting characteristics which, however, are quite different than methacrylate. Even a weak acetone solution in the trough is apt to wet the block face as soon as sectioning starts, and thus prevent proper flotation. However, pure water seems to work very well indeed, and a little acetone (less than 10%) can be added if it seems desirable to reduce the surface tension somewhat.

Araldite and Vestopal W blocks will be much harder than the usual methacrylate block, and for these, knives with more acute angles (less than 45°) are recommended. An acute knife makes possible a concave meniscus which helps prevent wetting the block face.

Araldite blocks, particularly when prepared in the manner of Moore (described in Chapter 4.7), may be excessively brittle, and chip as they are trimmed. Thus the block face may at first contain pits and cracks which fill with trough fluid and serve as centers for continual wetting. One has to be prepared to do much thick sectioning with the microtome to get through this zone before thin sectioning can begin. This difficulty is reduced or nonexistent if plasticizer is used, and is not a serious problem with some of the other embedding techniques.

It has already been emphasized that specimen contrast is intrinsically poor when these embedding media are used. In most situations very thin sections will prove to be useless. At least in starting work, sections with light gold colors should be emphasized. It may prove advisable to use 80–100 kv accelerating voltages for the most effective micrographs of such material.

This writer admits a rather limited experience in sectioning Araldite and Vestopal W material, but he frequently has thought he was obtaining beautiful sections only to discover later that they were permeated by a fine order chatter, or the section surfaces proved to be rough (Fig. 9(c)). Thus the blocks are apt to give a false impression that they are cutting very well indeed, when, in fact, the sections will be useless. No doubt one way to eliminate the chatter is to reduce the block size, but one hates to do this more than actually is necessary. This writer believes that some improvement results from the use of a plasticizer during embedding to soften the blocks. Perhaps it should be more generously used than the literature suggests.

5.19. Summary of Sectioning Technique

Woven into this and the previous chapter have been a number of precautions that must be taken for successful sectioning, as well as a discussion of difficulties that may be encountered. It seems wise to bring these together in the form of a summary or a check list that can be consulted in time of frustration. The beginner should be encouraged by the knowledge that ultrathin sectioning is a proven possibility, and that if he is having difficulties or is unsuccessful, it is probably because of some seemingly small flaw in his technique. He should take stock of every key step and not attempt shortcuts.

(1) Knife contamination. This is the most common difficulty, and the fingerprints which produce oily films are the worst offenders. Dirty towels, impure solvents, oily tools, contaminated containers, trough walls—all must be considered.

(2) Block face too large. The cutting forces deflect the knife edge with consequent digging and skipping, and often wide variation in section thickness along with the skipping. Even thin sections are apt to be useless because of fine-order chatter. This is second only to contamination as the principal cause of failure.

(3) Knife tilted too far forward. This favors digging and subsequent skipping. It is also apt to produce fine chatter which may not be apparent while sectioning is in progress, thus giving a false sense of security.

(4) Knife tilted too far backwards. There will be much skipping and occasional thick sections.

(5) Cutting speed too fast. This favors fine chatter which may not be apparent during the sectioning. Be sure you can follow the block with your eyes as it descends towards and across the knife.

(6) Unsteady cutting. The microtome must be cycled rhythmically; any interruption or break in this rhythm will require a period of readjustment before good uniform sectioning can begin again.

(7) A light touch. The microtome can be deformed enough to ruin quality sectioning simply by a firm grasp on the microtome handle.

(8) Meniscus. Too low a meniscus is apt to be the common fault. Back pressures may be created which will prevent the sections from flattening out on the fluid surface as they should. Too high a meniscus is apt to wet the block face which in turn will carry fluid to the knife face, which may result in contamination.

(9) Vibrations. The microtome table should be steady. The operator should not touch the table with hands, elbows, or feet while sectioning.

5.20. Sectioning with the Sjöstrand Microtome

This author has not had any real experience with the Sjöstrand ultramicrotome, but it is quite apparent that it is used somewhat differently than the Porter–Blum instrument. Since the Sjöstrand microtome has had such an influence on the development of European ultramicrotomy the difference is worth very particular consideration.

The actual cutting speed with the Sjöstrand ultramicrotome is several times faster than is commonly used in America with the Porter–Blum microtome, and which has been emphasized in this chapter. How is this possible? It seems evident that one pays for the speed with the coin of section size. Sjöstrand's sections are minute by American standards. Sjöstrand may run his microtome with its motor drive for half an hour or more, and end up putting well over 100 sections on one specimen grid. Our cutting in America produces sections of such a size that less than a dozen may pretty well cover a standard grid.

One might ask why the Sjöstrand microtome is not slowed down so that it could handle larger blocks? No doubt the thermal advance mechanism could be suitably adjusted to do this, but then one would be in the same dilemma the person finds himself in who tries to use a motor drive on the Porter–Blum microtome. The cycling becomes too slow, and thermal fluctuations become a serious problem, to say nothing of the irritating delay in getting sections. A virtue of the Porter–Blum microtome when turned by hand is the short cycle that can be established along with a very slow cutting speed. No doubt, this could be accomplished with a motor drive having a sufficiently complicated controlling mechanism.

Most of the new microtomes that are appearing in Europe today are being operated by thermal drives and their cycling is at a uniform rate. In most instances, cutting speeds are relatively high, of the same order as the Sjöstrand instrument. Allowances for this in terms of block size will have to be made by those who would operate them.

5.21. Picking Up Sections

Before sections can be examined in the electron microscope they must be mounted on suitable specimen grids and given additional support with a backing film of one sort or another. The various ways that sections can be supported are discussed in detail in the next chapter. Here we are concerned with getting the sections to the grid which may already have a supporting film upon it.

The simplest way to pick up sections is in most respects not the best way. But one need only take a screen in fine tipped forceps and bend its edge slightly so that it can be laid flat upon the surface of the fluid in the trough. It is simply laid across a ribbon or raft of sections, and then pulled away, or it may be pushed on down into the trough fluid, then turned over and finally removed. Sections stick vigorously to a dry screen or to a screen which has been coated with a supporting film. No further attachment is necessary other than simple contact. However, the screen must not be wet before contact so that it must not be tilted as one comes down over the sections, and this motion should not be made too slowly or condensation of water may occur.

The disadvantages of this simple method of picking up sections include: (1) the very real danger of damaging the knife edge, either by direct contact or by contamination; (2) the difficulty of orienting the sections very well on the screen, or assembling a group of sections in one place so that they all are put on a single screen; (3) the probability that any contaminants on the surface of the trough fluid will be scattered over the sections, thus dirtying them to the point where their usefulness is seriously impaired; (4) compression effects may show on freshly cut sections unless they have previously been expanded by xylene fumes as indicated in Chapter 5.14.

A variant of this method is to dip the screen first, and approach the sections from below. Sections are apt to be hard to center, though, if they are stuck to the knife or trough wall. This is, however, probably the best way to handle Araldite or Vestopal

W sections which are floated on water rather than on an acetone solution.

While these direct methods of picking up sections may be simplest, any saving of time or effort is apt to be nullified many times over subsequently. This may be evident when the screens are examined with the electron microscope only to reveal that much of each screen is blank, or that there are only a few decent sections, and that these are mostly dirty. A little more effort and care in picking up the sections in the first place should yield screens loaded with good clean sections. The method of direct pickup on grids is recommended, therefore, only for quick assay, perhaps to determine whether fixation has been adequate to make further careful study worthwhile.

A more elegant and flexible method of picking up sections involves transferring them to a dish as an intermediate step, where they can be studied under optimal conditions with a binocular dissecting microscope. The dish can contain something other than dilute acetone to assist in expanding the sections. We have found 50% alcohol to be quite effective in this connection. Also, this concentration of alcohol does not evaporate too fast which otherwise would require constant attention. Sections can be moved about in the dish and good ones collected together. "Staining" substances such as phosphotungstic acid or uranium salts can be added to the alcohol and will penetrate the sections. The fluid upon which the sections float can be given a flat meniscus, and therefore one can arrange lighting at will so that interference colors can be seen without difficulty. With care in handling, the sections should not become dirtied by surface contaminations. These, then, are the advantages of an intermediate step in picking up sections. The method is practical because there is a simple and almost foolproof way of removing sections from the trough and transferring them to the surface of another fluid.

To remove sections from the trough for transfer, one can take advantage of surface tension gradients. The only equipment needed is a "glass spatula." This is made by diagonally breaking an ordinary 1×3 in. microscope slide so that it has a blunt end, a little narrower than the trough. This is left standing and kept wet in a small beaker of the same 40% acetone which is used to fill the trough. Acetone, of course, evaporates fairly rapidly from the relatively shallow and broad-surfaced trough. If the tip of the wet spatula is quickly introduced into the trough (Fig. 16), the initial acetone concentration on its surface will be greater than that of the trough, and it will therefore have a lower surface tension. Sections will retreat from the spatula at first. The situation rapidly reverses, however, for the spatula has but a thin film of acetone upon its surface which evaporates quickly. When the acetone concentration of this film becomes less than that of the trough, the sections glide up the spatula upon this film following the surface tension gradient. They quickly, and without damage, accumulate on the surface of the spatula. Gentle nudging may be needed to free the part of the ribbon still attached to the knife edge. The whole operation takes much less time to accomplish than to describe. The surface of the trough fluid is automatically cleaned by this process too, which is an additional advantage of the method.

Since this method of picking up sections depends upon surface tension gradients, it is obvious that the acetone concentration in the trough must be reasonably large. If too much evaporates, the gradient may not be steep enough to be effective. Under most circumstances the simple solution is to replace the trough fluid with a new acetone solution from time to time. Sometimes one does not want to do this, for perhaps there are some good sections in the trough which should be salvaged first. The spatula then can be wet with pure water which will establish a steep gradient even when acetone concentrations in the trough are greatly reduced. It must also be understood that the spatula tip has to be truly wet by the acetone or water. It is prone to become hydrophobic with use, but it can be made hydrophilic by dipping it in alcohol, or perhaps by wiping it with an alcoholic solution, and then returning it to the acetone.

The container to which the sections are transferred should be about the shape of a Syracuse watch glass, although it need not be so large. This container must not be made of glass, however, for methacrylate sections adhere to dry glass with great tenacity and many sections inevitably would be lost by becoming irrevocably stuck to the side walls of the container. In our experience "Bakelite" plastic containers are best for there is only a mild tendency for the sections to adhere to this material. Very satisfactory containers of this material are commonly available in hardware, furniture, and "dime" stores as "caster cups," made to be put under the legs of chairs, etc., to protect floor coverings. These are available almost in the pattern of a Syracuse watch glass.

The plastic dish is filled exactly to its lip with the flotation solution. As indicated above, 50% alcohol seems very satisfactory. The meniscus should be flat. To transfer sections to this surface, the tip of the spatula with sections upon it is dipped into the alcohol. A little alcohol is taken out of the dish with a medicine dropper pipette and the surface of the spatula is flushed with this. Naturally the same amount of alcohol is returned to the dish as was taken out, so that the fluid level remains constant. The sections are not damaged by being floated down the spatula and onto the alcohol surface. They will quickly leave the center of the dish and accumulate around its rim, making light contact with the rim. The alcohol must not be allowed to evaporate far, however, or the sections may become stuck to the exposed side walls.

Sections collected at the rim of a plastic dish may be examined at one's leisure, and under good conditions of illumination. The binocular dissecting microscope used for this purpose should be tilted so that the interference colors can be seen to best advantage. Small clusters of sections can be moved in the dish by using a wet toothpick that has its tip fractured and bent so that the point can come up from the alcohol below the section to be moved. In this way good sections can be grouped together, and nearly complete sheets of sections can be aggregated so that the entire specimen screen is pretty well covered with promising material.

Sections are picked up close to the rim of the dish, either by touching them with a supporting grid and then pulling it away, in the manner described for the direct pickup from the trough, or by coming down on top of the sections and then following through so that the grid and forceps tips are submerged in the alcohol. The grid is then turned over while under the alcohol, and lifted out of the solution. Its lower surface is touched to a piece of filter paper, and then it is placed on a glass slide to dry.

By far the cheapest mounting screens for electron microscopy are made by punching sheets of a copper grid material which is not the same on its two sides ("Lektromesh"). On one side, each wire is convex, while on the other side it is flat. This low cost material is perfectly satisfactory for most work. For theoretical reasons it seems best to place the flat side in contact with supporting films and sections, for this would provide the greatest possible area of contact and favor adherence. In practice, however, it does not make much difference.

The more expensive "Athene" type grids, made by a different process, have one side polished and the other matte. The unpolished side should be in contact with supporting films or sections for adequate adherence.

5.22. "Staining" for Electron Microscopy

When one speaks of "staining" for electron microscopy, he is, of course, thinking of the addition of heavy atoms which scatter electrons readily and so enhance the final contrast of the picture. Of course, osmium tetroxide has very important staining properties associated with its qualities as a fixative. Indeed, electron microscopists, more or less seriously, have said that the micrographs that we finally study represent little more than patterns of osmium deposition. One has only to consider the low contrast of material fixed with formaldehyde to realize that there is a large measure of truth in this statement. Fortunately osmium tetroxide seems to combine in such a variety of ways with many things that it quite effectively serves as a general purpose stain.

There have been continued efforts to "stain" structures that are not particularly affected by osmium tetroxide, and even to boost the contrast of those which are. Until recently, however, only a single agent really had proved effective. This was phosphotungstic acid which originally was introduced by Hall *et al.*

(1945) in their studies of collagen. This is a most curious compound in that 24 atoms of tungsten are combined in each molecule. Thus, every molecule that can be attached to a binding site enhances electron density tremendously. Ordinary compounds containing one or two heavy atoms that might be similarly attached cannot possibly be as effective.

The work with collagen suggests that phosphotungstic acid has quite specific binding sites in proteins. However, presumably all proteins have such sites so that in effect, this substance serves in some degree as a "universal" stain for proteinaceous materials. It is particularly effective in tissue work in staining collagen and elastin, as well as cytoplasmic fibrillar structures.

Postfixation staining with phosphotungstic acid has achieved a bad reputation in some circles because, undoubtedly, this substance can cause much tissue damage if not handled with some finesse. Phosphotungstic acid in aqueous solution has a low pH. This acidity, perhaps, causes most of the damage. To avoid this, some investigators have added phosphotungstic acid to the absolute alcohol used during the dehydration of the specimen (Figs. 9 and 10). More recently it has been realized that materials dissolved in water will penetrate sections in spite of the embedding medium. Phosphotungstic acid can be added to the acetone solutions used in the trough to spread sections as they are cut. Or better yet, sections can be transferred as previously indicated to the surface of dilute alcohol solutions containing phosphotungstic acid (Figs. 2 and 21). Delicate staining, which apparently has no harmful effect on the section, can be obtained with concentrations of phosphotungstic acid as low as 0.1% in 50% alcohol. Staining time need not be longer than 10 min. In these concentrations no special washing is needed to get rid of the excess phosphotungstic acid. Even 1% solutions may not require washing, although the investigator should look out for granular deposits. Much stronger solutions, however, will require a wash. This can be done most simply just after sections have been picked up on screens. They are dipped momentarily in clean, 50% alcohol before drying.

A survey of the literature dealing with ultrathin sectioning no doubt would disclose many attempts to use heavy metal salts to enhance contrast. In general, though, investigators have not been particularly concerned with the technical problems, and these aspects usually were not controlled adequately to demonstrate convincingly that there were important and desirable effects. Watson (1958), however, has studied carefully a group of compounds, three of which show great promise as generally useful electron stains. His bibliography will acquaint the interested reader with the little useful information of the past.

Watson advocates the use of uranyl acetate as a general purpose electron stain which can be used easily. Although the molecule is simple, the uranium atom is sufficiently heavy to give significant electron scattering wherever it may be attached. He floats sections, after mounting, face down for 1–2 hr upon saturated solutions. They are then rinsed briefly. He presents some evidence of selective specificity of this compound. Sodium uranate was not found to be as effective as the uranyl ion.

Yet another general purpose electron stain is alkalized lead acetate. Watson (1958) prepares this by starting with about 2 ml of a saturated solution of lead acetate, and adding 40% NaOH, a drop at a time. At first the precipitate formed by each addition is dissolved by stirring. The solution is ready to use when a faint opalescence remains. The solution is not stable, and it is used at once. It stains in about 5 min, and long exposures are contra-indicated as they may allow extractions of tissue components.

Lead hydroxide, prepared in a particular manner by Watson (1958) has been found to be a most vigorous general purpose stain. Watson, in fact, believes it to be the most effective of all the various compounds he has studied. He dissolves 8.26 gm of lead acetate in 15 ml of distilled water to produce a nearly saturated solution. The lead salt has to be ground under the water with a glass rod. This is put into a centrifuge tube, and 3.2 ml of 40% NaOH is blown rapidly into it from a pipette which produces a dense precipitate. After stirring, the precipitate is sedimented. The supernatant is decanted and discarded. The precipitate is washed by resuspending it in the original volume of fluid, centrifuging, and decanting. It is resuspended once again and, after centrifuging, the supernatant becomes the staining solution. The solution is reasonably stable if stored at room temperature (not refrigerated). A little of the precipitate may be included with it to maintain saturation.

General staining becomes increasingly intense during the first 30 min (Fig. 18). The stain is also specific for what are thought to be glycogen particles, but then exposure should be only 1–2 min.

There is one unfortunate difficulty with lead hydroxide. The carbon dioxide of the air is sufficient to form completely insoluble and hydrophobic crystals of lead carbonates which quickly form a scum on the surface of the solution. These crystals easily dirty a section to the point of its becoming useless. This can be minimized by storing the solution in a plastic syringe, and extruding isolated drops with fresh surfaces as needed. Peachey (1959) details how a "disposable" plastic syringe can be modified easily for this purpose. The syringe is mounted vertically, and a small well accommodating a grid is cut in the tip of the nipple. The original needle cap is used as a dust cover for the staining well, and also as a CO_2 trap after packing it with a sodium hydroxide pellet and some anhydrous calcium chloride. Washing the grids with sections after staining should be done with freshly boiled demineralized or distilled water.

Watson (1958) quite rightly warns about the pharmacological dangers of the very finely divided lead salt crystals that may be spread about unless suitable care and precautions are observed in cleaning containers and overflow basins.

Watson (1958) also has studied several other compounds including phosphomolybdic acid, ammonium molybdate, and barium hydroxide, but these have little general interest. The

FIG. 18. Rat kidney "stained" after sectioning for 10 min with Watson's lead hydroxide solution. (a) The wall of a glomerular capillary. The plasma membranes of epithelial cells are very intensely stained. Precipitated blood plasma and intracellular protein show unusual density. (b) Apical ends of cells of a proximal tubule. Some tiny cytoplasmic granules are intensely stained. (c) Basal ends of cells of a distal tubule. The infolded plasma membranes are seen vividly. Mitochondria are less affected.

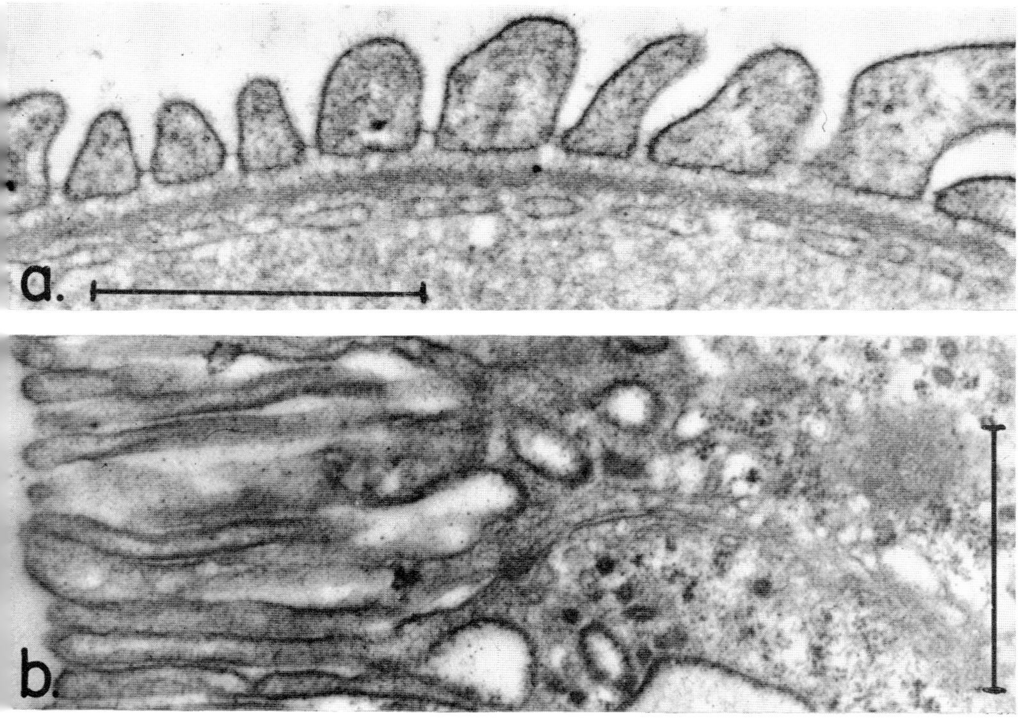
a.
b.
c.

latter, however, differentiates a particular component of the Golgi apparatus.

When tissue is embedded in Araldite or Vestopal there are special difficulties with contrast as has been emphasized in Chapter 4. More than ever, such tissues are apt to require heavy metal staining in order to allow adequate visualization of components. In general, it would seem that people have not had much success staining such sections by floating them on fluid surfaces after they have been cut. Most investigators, therefore, have soaked tissue blocks in solutions of staining agents before embedding. To a limited extent this is successful. Brody (1959) has demonstrated, however, that heat may help drive staining agents into Araldite sections. He achieved good staining of difficult material (epidermis) using saturated aqueous solutions of uranyl acetate at 40–70° C for periods up to 2 hr. Swelling agents, such as xylol vapors, have been used by other investigators, apparently with some success.

5.23 Thick Sections for Conventional Microscopy

The electron microscopist is frequently in the position of wishing to observe his material with conventional microscopy. His purposes may range from simply needing an identification of a structure as being present at the surface of the block, to highly refined microscopy, perhaps utilizing histochemical techniques. He may wish to combine phase microscopy with electron microscopy.

For the simple identification of structures large enough to be characterized with low power objectives, very crude techniques may suffice. Even freehand sections may be cut from a block face with a sharp razor blade and mounted in glycerin or immersion oil under a cover slip, and then be examined with a 16-mm objective with the condenser diaphragm nearly closed. Ordinarily, enough will be seen to indicate the tissue pattern, so that, for instance, glandular alveoli might be seen, a small arteriole would be apparent, or a glomerulus identified.

The next stage of refinement is to cut suitably thick sections with the microtome, which are then picked up on pieces of glass (suitably small pieces of cover slip may serve to pick up sections directly from the fluid surface of the trough, or sections may be transferred to standard slides). If the glass is clean, the sections will adhere vigorously to the glass and no further cementing will be necessary no matter what is to be done to them subsequently. Temporary mounts of such sections can be made and studied with phase microscopy, either before or after extracting the methacrylate. Thus, in just a few minutes one can obtain much information about the contents of a section, and be in a position to continue wth more refined work.

To some extent staining reactions can be carried on even without removing the methacrylate. Usually, though, more vivid and reproducible results will follow extraction of the embedding medium. Then almost any staining reaction will work reliably. Bencosme and associates (1959) have emphasized the usefulness of toluidine blue in this connection, since metachromasia may emphasize differentiation. They use acetone to dissolve the methacrylate and to expedite hydration of the section. After staining with the aqueous dye they often can see enough detail in a temporary aqueous preparation to satisfy their needs. Or a permanent mount can be made in a matter of minutes by dehydration with acetone and mounting conventionally. A much more elaborate polychrome staining method utilizing a gallocyanin–chromium mixture followed by phloxin B, has been proposed by Runge *et al.* (1958).

The complaint of the conventional microscopist in viewing even well-stained sections of the electron microscopist is that they tend to lack dramatic contrast. To be sure, the osmium tetroxide fixation creates a gray tone, which, however, may be somewhat "bleached" by prolonged soaking in dioxane. More importantly, though, the excellent preservation of the electron microscopist preserves so much material so uniformly in place that there are not the major discontinuities which exaggerate contrast in conventional microscopy. So much material remains in place to be stained that it is hard not to overstain a conventionally thick section with an apparent loss of detail. Actually the conventional light microscopist can see all of the detail that he is used to and more if he has a well stained section about 1μ thick. If he appreciates that he is seeing a better preserved tissue that he has ever looked at before, his experience will be more meaningful.

Literature Cited

Bachmann, A., and Sitte, P. (1958). *Program (Abstr.) Fourth Intern. Conf. Electron Microscopy, Berlin,* p. 106.

Bencosme, S. A., Stone, R. S., Latta, H., and Madden, S. C. (1959). *J. Biophys. Biochem. Cytol.* **5,** 508.

Bretschneider, L. H. (1952). *Intern. Rev. Cytol.* **1,** 305.

Brody, I. (1959). *J. Ultrastructure Research* **2,** 482.

Claude, A. (1948). *Harvey Lectures, Ser.* **43,** 121.

Fernández-Morán, H. (1956). *J. Biophys. Biochem. Cytol.* **2** (Suppl.), p. 29.

Fullam, E. F., and Gessler, A. E. (1946). *Rev. Sci. Instr.* **17,** 23.

Gettner, M. E., and Hillier, J. (1950). *J. Appl. Phys.* **21,** 68.

Gettner, M. E., and Ornstein, L. (1956). *In* "Physical Techniques in Biological Research" (G. Oster and A. W. Pollister, eds.), Vol. III, Chapter 13. Academic Press, New York.

Hall, C. E., Jakus, M. E., and Schmitt, F. O. (1945). *J. Appl. Phys.* **16,** 459.

Huxley, H. E. (1957). *J. Biophys. Biochem. Cytol.* **3,** 631.

Latta, H., and Hartmann, J. F. (1950). *Proc. Soc. Exptl. Biol. Med.* **74,** 436.

Morgan, C., Moore, D. H., and Rose, H. M. (1956). *J. Biophys. Biochem. Cytol.* **2** (Suppl.), p. 21.

Newman, S. B., Borysko, E., and Swerdlow, M. (1949). *J. Research Natl. Bur. Standards* **43,** 183.

O'Brien, H. C., and McKinley, G. M. (1943). *Science* **98,** 455.

Peachey, L. D. (1958). *J. Biophys. Biochem. Cytol.* **4,** 233.

Peachey, L. D. (1959). *J. Biophys. Biochem. Cytol.* **5,** 511.

Pease, D. C., and Baker, R. F. (1948). *Proc. Soc. Exptl. Biol. Med.* **57,** 470.

Porter, K. R., and Blum, J. (1953). *Anat. Record* **117,** 685.

Richards, G. A., Anderson, T. F., and Hance, T. R. (1942). *Proc. Soc. Exptl. Biol. Med.* **51,** 148.

Runge, W. J., Vernier, R. L., and Hartmann, J. F. (1958). *J. Biophys. Biochem. Cytol.* **4,** 327.

Satir, P. G., and Peachey, L. D. (1958). *J. Biophys. Biochem. Cytol.* **4,** 345.

Sjöstrand, F. (1943). *Nature* **151,** 725.

Sjöstrand, F. (1953). *Experientia* **9,** 114.

von Ardenne, M. (1939). *Z. wiss. Mikroskop.* **56,** 8.

Watson, M. L. (1958). *J. Biophys. Biochem. Cytol.* **4,** 475 and 727.

Weiner, S. (1959). *J. Biophys. Biochem. Cytol.* **5,** 175.

Williams, R. C., and Kallman, F. (1955). *J. Biophys. Biochem. Cytol.* **1,** 301.

6. Section Mounting

6.1. Introductory Remarks

Unless supporting grids of extraordinarily fine mesh are used, methacrylate sections require an additional supporting film of some sort, for methacrylate is softened and partially decomposed in the electron beam. It has been recognized since the earliest days of electron microscopy that supporting films are necessary evils. Insofar as the film possesses mass, it scatters electrons and reduces contrast. One can imagine the situation of a specimen having the same density as the film. If it were embedded in the film it would then be quite invisible. Material fixed in formalin approaches this condition, particularly if embedded in Araldite or Vestopal W. Thus, from the very beginning, efforts have been directed towards making the thinnest possible supporting films of materials consisting of light atoms. This technology had progressed considerably before ultrathin sectioning ever got started; for films are used by bacteriologists to support bacteria and viruses, and by chemists and physicists to support all sorts of particulate matter.

It was natural that investigators turned to plastics in searching for good supporting films. The requirements were for materials which could be dissolved simply, and which would form tough films when the solvent was allowed to evaporate. Parlodion and Formvar were used early for such films and remain the most satisfactory plastics today, although superior films of pure carbon supplant these old standbys to a considerable extent. Carbon films are formed by the deposition of vaporized carbon *in vacuo,* a technique introduced by Bradley (1954). Pure carbon films (or carbon stabilized plastic films) are so generally used now that a high vacuum evaporating apparatus has to be regarded as an essential item of equipment in an electron microscope laboratory in spite of its considerable cost.

Although simple Parlodion and Formvar films should not be used anymore for quality work, they are often used in combination with a carbon deposit which stabilizes them, and they are used as a possible step in preparing carbon films since they can be dissolved away after carbon has been deposited, leaving a pure carbon film behind. Thus, the beginner has to know how to prepare such films, and this is discussed first in the succeeding pages before considering carbon.

6.2. Parlodion Films

Parlodion is the trade name of a nitrocellulose plastic manufactured by the Mallinckrodt Chemical Works, St. Louis. It has perhaps the highest viscosity of any nitrocellulose which gives it an advantage over its relatives of forming the toughest possible films with the least mass of material.

Parlodion films are easily prepared. Stock solutions of 0.5% (or less) Parlodion are made up in amyl acetate (or sometimes ethyl acetate). It takes about two days to dissolve. During this time, the plastic swells in its solvent and becomes deceptively invisible long before true solution occurs, so it should be stirred from time to time.

Thin Parlodion films are formed on standard 1 × 3 in. microscope slides, and then they are stripped away from the glass. The microscope slides should not be too carefully cleaned for then the film may become too adherent to the glass to strip easily if at all. Thus, new microscope slides should simply be wiped with a clean towel to remove obvious lint and dust. Fingerprints on the glass surface, of course, should be avoided. A slide is dipped into the Parlodion solution and held there for a few seconds. This gives time to complete any solvent activity that might involve contaminating layers at the surface of the glass. The slide is then withdrawn and drained as rapidly as possible. The flat plane of the glass should be held vertically, the long axis of the slide should be held obliquely so that excess solution drains quickly across the short axis of the glass, and then runs down one edge of the slide to a corner. Since amyl acetate evaporates relatively slowly, drainage essentially is completed before the viscosity of the remaining solution begins to impede it. Thus the deposited film will be very nearly of uniform thickness except at the draining edge of the slide. This technique of rapid draining also allows the standardization of film thickness, and the thickness is quite directly related to the original concentration of Parlodion.

After the film is dried on the glass surface it is floated off the slide on a clean water surface. For this one needs a good sized glass bowl. Circular glass baking dishes about 10 in. in diameter serve well (Fig. 19(a)). One should take certain precautions to get a clean water surface. The glass bowl itself should be clean, of course. Tap water can be used, at least in some cities. If the water tends to be supersaturated with air, it is best to let it run from the faucet for a few minutes until only a few bubbles appear as the bowl is filled. The bowl should be filled to overflowing and all bubbles allowed to rise to the surface. Then excess water is spilled over the edge and the clean water surface is ready to use.

The next step is to take a moderately sharp object such as a forcep's tip and make a rectangular score mark close to the edges of the slide, around the area of film that is to be removed.

Now a mystical ritual should be performed. The slide is held near the open mouth and one vigorously breathes on it to "frost" its surface with condensed water vapor. While still frosted, the slide should be dipped into the water at an angle of about 30° from the horizontal. As the dip starts, observe the end of the slide and be sure the film is loosening properly so that it floats off on the surface of the water as the slide sinks down into it (Fig. 19(a). Have a toothpick handy which can be wet, and the film pushed away from one's fingers after it has been floated.

A suitably thin film will be hardly visible, and certainly will not show interference colors. It can be seen only by looking at it with reflected light. A broad fluorescent light source is excellent for this. It is usually convenient to have placed a specimen grid at each end of the film before stripping it. These then serve as markers.

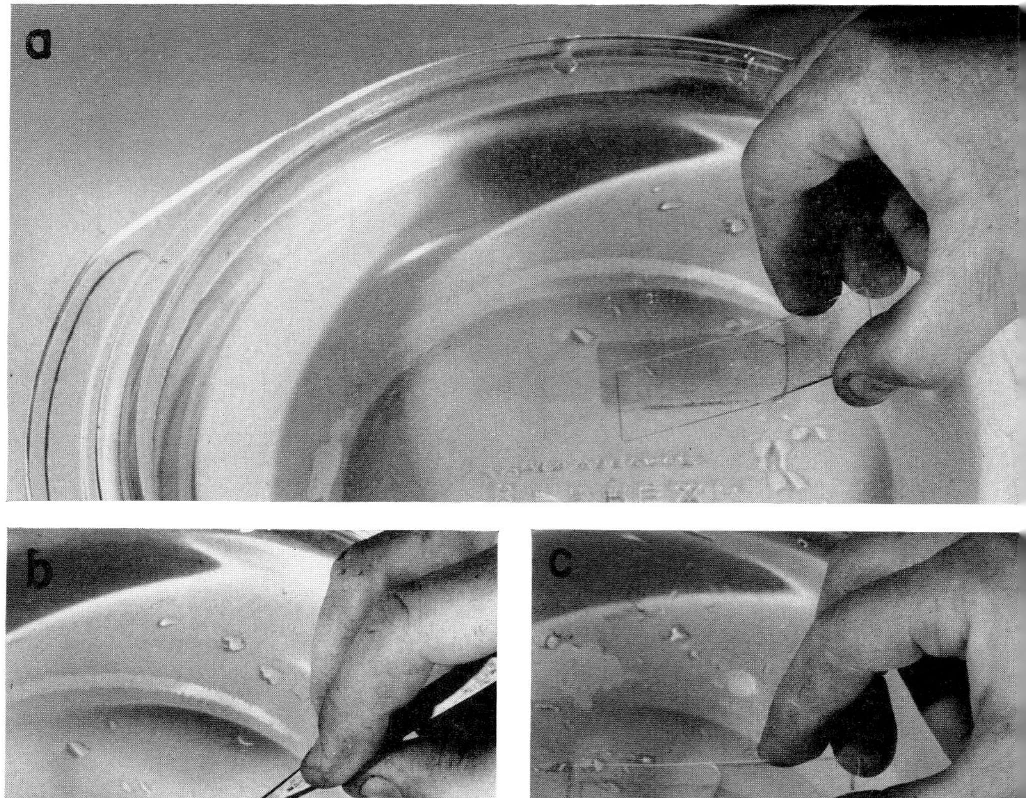
a
b
c
d
e

Specimen screens are placed in rows on the floating film (Fig. 19(b)). Two precautions should be observed. If the screens have been punched in their manufacture, they have a slight burr around their edges so that one side is convex and the other concave. It is important that the convex side be the side in contact with the Parlodion film. Otherwise the film may stretch from the edges of the burr, and make only occasional contact with the screen. The membranes then invariably rupture and the screens are quite useless. In fact it is wise to bend the screens slightly to increase the convexity in contact with the film. Also, rows of screens should not be placed too close to one another for there must be reasonably large expanses of exposed film to make contact with glass during the pickup operation described next.

Before proceeding further, each floating screen should be tapped gently with a needle so that the film is depressed. This assures a good bond between the screen and film. Then one is ready to pick up the screens from the water surface. This is done by taking a dry microscope slide and, while holding it nearly horizontally, coming down on top of the floating film with its screens (Fig. 19(c)). This motion should be fairly fast to prevent condensation of water on the glass, which then would not allow the glass and film to adhere to one another properly. The slide is pushed right down through the water surface and

---

FIG. 19. (a) Parlodion or Formvar films, cast on the surface of ordinary microscope slides, can be floated off on a water surface as indicated in (a) (the film was smoked to increase visibility). To do this the edges of the film first are scored, and then moisture is condensed on its surface by exhaling upon it. The slide is held at about a 30° angle as it is pushed down into the water. The film should float away freely from the glass as shown here. A fairly large glass "baking dish" should be used for this purpose. It should be nearly filled with water, and the water surface must be clean.

While the film is floating on the water surface, supporting screens can be laid upon it (b). They should not be spaced too closely together. Each should be tapped with forceps tips to assure good adherence. If one face tends to be convex, this should be the face in contact with the film.

After the screens are in place, a dry microscope slide is brought down on top of the film at about a 30° angle (c) and without delay it is pushed down into the water carrying the film before it. The slide is then turned over while under the water (d) and removed face up with the film and screens on top. Excess water is removed with a bit of filter paper (e) and the screens are ready to use after drying.

when it is totally submerged it can be turned over (Fig. 19(d)) so that the film with the screens are on its uppermost surface. There is no danger of the film or the screens floating away if the indicated precautions have been observed. The slide with the film overlying the grids next is lifted out of the water and drained. Large droplets of water in the vicinity of screens should be removed with the edge of a piece of filter paper (Fig. 19(e)). As soon as the slide has dried the screens are ready to use, each now being coated with a thin film of Parlodion.

If high quality screens are being used which are very flat, or if screens with unduly large holes are being employed, surface tension may pull the Parlodion films down to the glass surface where they may eventually stick as drying is completed. Thus the supporting films may be lost as the screens dry upon the glass slide. This usually can be prevented if a piece of paper towelling is used instead of the glass slide to pick up the screens from the water surface. However, it is then almost necessary to transfer the screens to a dry towel or to a glass slide to complete the drying process. Paper towels having a fairly hard surface should be chosen for this since they are less apt to shed strands of lint.

As the beginner's skill increases and he wishes to make high resolution pictures, he will want thinner and thinner supporting films, even to the point that the films are so thin that they usually break. Extra thin films are made quite simply by using more dilute solutions of Parlodion to cast them in the first place. Since "dry" Parlodion is apt to contain some water it is not easy to weigh accurately. This author has had more success preparing very thin films by diluting standard solutions of Parlodion rather than by trying to prepare reproducibly dilute solutions directly. With each successive dilution the character of test films can be assayed in the electron microscope.

Parlodion solutions in amyl acetate are reasonably stable so that they can be kept for a month or more without taking precautions. However, there is a gradual decomposition so that eventually one realizes that the films being prepared are not as strong as they should be, and then it is time to replace the solution.

As one further precaution, the reader might be warned that this writer has encountered bad lots of amyl acetate from which we could not get satisfactory films until they were redistilled. We obtained substantial quantities of more volatile contaminating substances, as well as a residuum with a substantially higher boiling point than amyl acetate.

Films of Parlodion also can be made by dropping 1 or 2 drops of 0.5% amyl acetate solution on a large clean water surface. Surface tension spreads the solution before evaporation sets the film. Films made in this way look beautiful, but they do not seem to be as strong proportionately as films cast on glass, and therefore this method is not recommended.

6.3. Formvar Films

Formvar is the trade name of a polyvinyl formal plastic manufactured by the Shawinigan Products Co., New York.

Formvar supporting films are generally regarded as being slightly stronger than Parlodion ones, and therefore have been the films of choice in many laboratories in spite of the fact that they are a little more difficult to prepare, and stock solutions do not keep well. If one is going to use carbon to stabilize a plastic film (Chapter 6.7) or uses a plastic film simply as a substrate upon which to build a carbon film, differences are negligible and the more easily handled Parlodion seems preferable.

In principle, Formvar films are made just like Parlodion ones. However, there is no solvent for Formvar which behaves as nicely as amyl acetate for Parlodion. Most commonly, ethylene dichloride has been used as the solvent for Formvar, and the stock solution is usually 0.2–0.5%. In this writer's experience, film quality may deteriorate noticeably within days after a fresh stock solution in this solvent has been prepared, although its life may be lengthened by storing in a refrigerator. Chloroform is used in some laboratories as the solvent of choice, and these solutions seem to be more durable. Chloroform evaporates so rapidly, however, that it is difficult to drain the slide upon which the film is being cast in a uniform and reproducible manner, and the cooling which occurs with evaporation may cause water condensation. One can minimize this difficulty by draining the slide in an atmosphere nearly saturated with chloroform vapor. If the dipping solution fills only the bottom of a tall container, the upper part of the container will be filled with chloroform fumes, and the slide, after dipping, can be held there until it is drained. It should then be withdrawn sufficiently slowly so that evaporation does not chill the slide to the point where water condenses upon the still liquid film.

After the Formvar film has dried upon the glass slide it is stripped just as a Parlodion film would be. However, it is commonly somewhat more difficult to remove. Again the beginner should be advised not to clean the microscope slide too thoroughly; for this favors undesirably strong adhesion between glass and plastic.

6.4. Preparing Film Nets

Because supporting films inevitably reduce contrast and limit resolution, efforts have been made to avoid using them. Two approaches obviously are possible. One is to use finer and finer mesh supporting grids until the holes are so small that naked methacrylate sections stretched across them survive. Modern technology in making grids allows this to be done, and grids of almost any mesh size can be ordered.

It is also possible to make films of Parlodion or Formvar which are in the form of fine nets rather than being continuous sheets. These nets, when stretched over conventional screens, and particularly after they have been stabilized with carbon (see below), provide a simple solution. Such nets, as described by Sjöstrand (1957), are easy to prepare by condensing moisture droplets on the surface of the film just before it hardens. A glass slide is dipped in a 2% solution of Formvar in ethylene dichloride. It is then drained for about 20 min. in an atmosphere saturated with this solvent. Finally, one can breathe on the surface so that it "frosts," or it is easy to blow through a flask of hot water and create an atmosphere supersaturated with water which will condense on the glass surface. The nets so formed are floated off the slide, and stabilized with carbon (see below). The average size of the holes so formed is controllable within limits by varying the timing and degree of condensation. More simply the glass slide may be dried at once while observing its surface with reflected light. As soon as interference colors begin to appear, it is time to begin lightly "frosting" the film by breathing on it. As the drying proceeds across the slide, this is continued. The condensed moisture droplets create the holes. Very coarse nets can be prepared if grids previously coated with this sort of Formvar film are heavily bombarded in the electron beam of the microscope before mounting a specimen. The holes can be observed growing during the bombardment, and their size can be controlled at will. Finally the net can be stabilized by evaporating carbon upon it.

The method is particularly useful for studying very minute objects that are fairly common in the sections so that there is a good statistical chance of finding the object nicely centered in a hole in the mesh. Since the useful holes are much smaller than a single cell, this means that organelles or inclusions such as mitochondria, cilia, or specific granules are likely candidates for this treatment. However, there is an additional hazard in that unsupported methacrylate when bombarded by electrons is softened so that movements and displacements should be anticipated in unsupported material, particularly when very small dimensions are being considered. These can often be observed visually in the fluorescent screen of an electron microscope. The amount of softening, of course, depends upon the degree of heating and this is minimal in very thin sections.

6.5. The Mechanics of Carbon Evaporation

An ideal supporting film must be strong, stable, and thin. It should also be made up of light atoms, be elastic, and adhere tenaciously to metal supporting grids. Carbon falls short of the ideal in the last two of these characteristics, so that pure carbon films are, in fact, somewhat difficult to handle. They are extremely brittle and do not adhere to grids at all well, but they do possess a rock-like stability even when they are so thin as to have excellent electron transparency. Furthermore, they can be used as stabilizing support for plastic films that otherwise would be too delicate to be useful. It should be realized that pure plastic films (Parlodion, Formvar, or methacrylate itself) tend to drift or creep in the electron beam, and it is this movement during photography that often destroys resolution. Thus, carbon used in various ways has proved to be a boon to electron microscopists. The carbon evaporating apparatus has become an essential piece of equipment in every laboratory doing section work. The apparatus ordinarily is the same as that used for metal shadowing which has its major application in other areas of electron microscopy.

Film deposition by vacuum evaporation has come to have a number of industrial applications, most notably the "coating" of optical lenses and prisms to minimize surface reflections. Thus, there are commercially available both large and small equipment packages for this which are easily adaptable to carbon deposition. These units contain a pumping system consisting of a mechanical forepump and a high-capacity diffusion pump. They contain suitable valving systems arranged to protect the diffusion pump, and they have gauges which inform the operator of the vacuum. All of these features are desirable, but they are costly. Satisfactory carbon films can be produced in what the physicist would regard as a very moderate vacuum. The vacuum produced by a high quality mechanical pump actually is all that is needed. If necessary to economize, one can dispense with gauges and valves and use the mechanical pump directly to evacuate the chamber in which the carbonizing is performed.

Carbon is evaporated by heating it to a brilliant white incandescence in the vacuum. In the absence of oxygen, atomic carbon flies in all directions until colliding with some part of the apparatus or with any residual gas molecules in the chamber. Thus, films are built up on surfaces one atom at a time. Since carbon does not melt, and sublimes only at 3560° C, it is not affected by any heat generated by electron bombardment in the electron microscope.

Carbon has sufficient electrical resistance so that thin rods of it can be heated until they vaporize by passing moderate currents through them. In practice, one or two carbon electrode rods are given tips of small dimensions (Figs. 20(a) and (b)). A spring holds one rod against the other, and high amperage, low voltage current is passed through the system. If short tips about 1 mm in diameter are used it will require about 40–60 amps of 12-volt current (or the equivalent at somewhat higher or lower voltage) to evaporate the carbon. The current is supplied by a heavy duty transformer (1 kva capacity), and the secondary leads, of course, must be of very heavy wire or cable so that they do not become unduly heated.

The actual design of the apparatus to hold the carbon rods is not critical. A commercially available apparatus is shown in Fig. 20(a). There also are various ways of preparing the tips of the carbon rods that are to be vaporized. Most simply, one carbon rod can be sharpened in an ordinary pencil sharpener, and its sharpened tip butted against a flat end filed on the other rod. With such a system one must be careful not to place the specimens to be carbonized in the shadow of the broad-ended carbon. It must be realized that the carbon atoms will travel in straight lines from the point of vaporization, and anything intervening between this spot and the specimen will shield the specimen so that it does not get properly coated.

A better way to make carbon tips is to turn short, thin shafts of the type illustrated in Figs. 20(a) and (b). A tool to do this can be purchased commercially or can be made in a machine shop without much difficulty, using pieces of tool steel or carbide for the actual cutting edges. A commercially available tool is illustrated in Figs. 20(c) and (d). In some laboratories carbon rods are tipped in this way on a lathe. In this author's laboratory short lengths of carbon rod are mounted in the chuck of an electric hand drill, and spun in the commercial sharpening tool. Twisting the rods by hand will suffice, however, with such a tool.

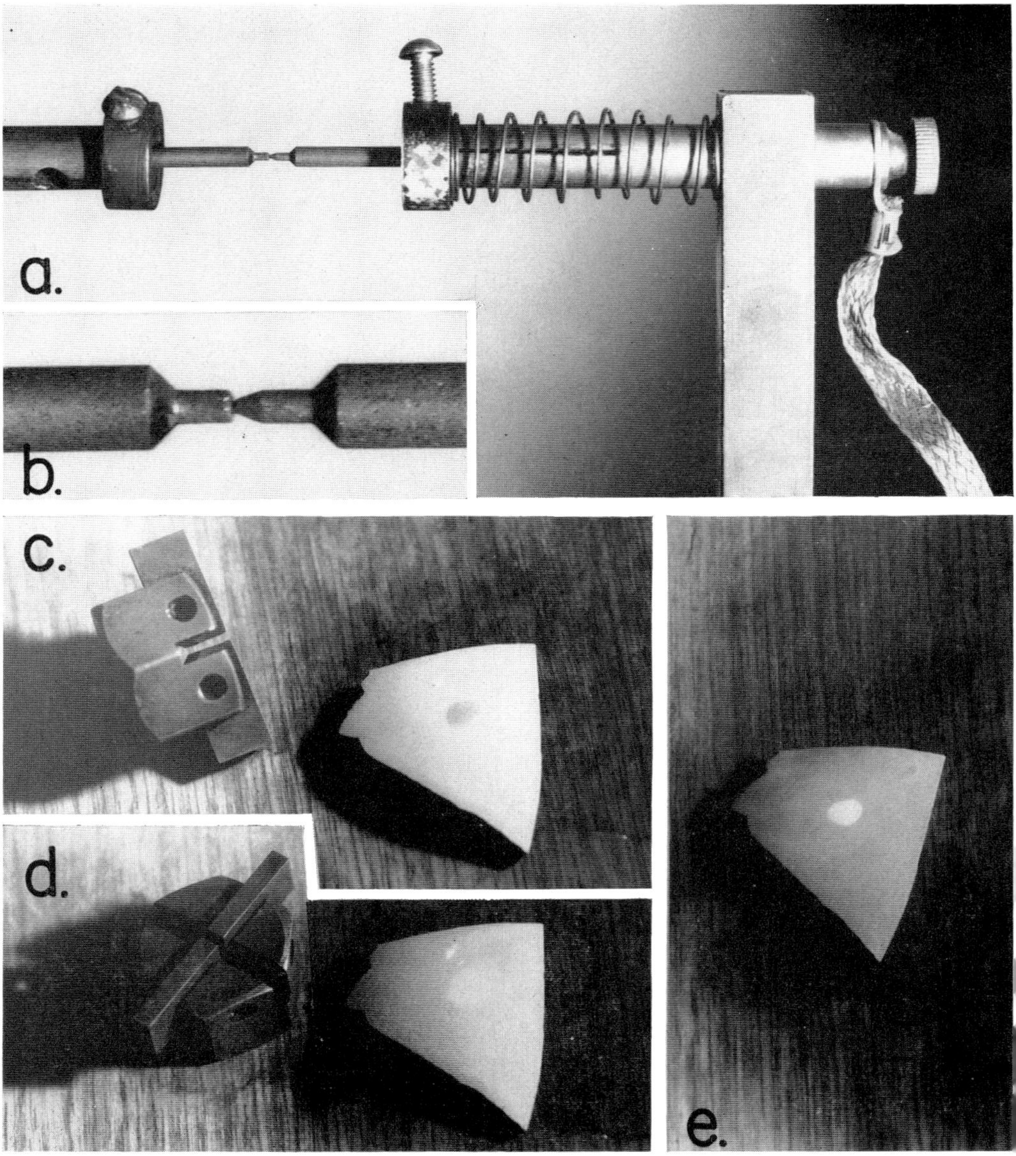

FIG. 20 (a) A commercially available holder for the carbon rods used in carbon evaporation. One ⅛-in. rod is fixed, the other is pushed into it by the spring-driven sliding carriage. (b) An enlargement of the tips of the carbon rods. These

A nearly standardized product can be obtained with any of these more refined methods. Finally, one carbon tip is sharpened with a fine file and the other flattened as illustrated in Fig. 20(b). These tips are lined up carefully in the carbonizing apparatus. When evaporization starts, then, a crater begins to form in the central region of the flattened tip, and the tips evaporate fairly uniformly. If the alignment has been sloppy, the crater may break through the edge of the flattened tip and the tips be deflected. Thus, one tip may glide past the other and effective contact be lost. Then one has the frustrating experience of being unable to continue with the evaporation until the chamber has been opened, the tips readjusted and the whole system reevacuated. Rarely is the adage of "haste making waste" more applicable.

Ordinarily the object or objects to be coated with carbon should be placed 5 or 6 in. away from the source to avoid excessive heat.

When the system has been suitably evacuated, carbonization can begin. The current input probably will be controlled with a variable transformer in the primary circuit of the transformer which is to supply the low voltage, high amperage current. Thus, as one turns a handle, the current increases. It is perhaps best to begin slowly until the carbon tips start glowing, then the operator knows that he has electrical contact. Next he should increase the current rapidly until a few sparks begin to fly in the

---

are shaped by the commercially available sharpener visible to the left in (c) and (d). Note that one tip is squared and the other pointed. It is important to center these properly in preparation for use. (c) This includes, besides the carbon rod sharpener, a piece of broken porcelain with a drop of vacuum-pump oil in its center. This serves as an indicator of the amount of carbon deposited. The porcelain is placed about the same distance away from the carbon points as is the specimen. (d) When carbon is to be evaporated directly upon a naked section overlying a support screen, or when carbon is to be used to stabilize a plastic supporting film, the carbon layer should be very thin, almost the least amount detectable on the porcelain indicator. The carbon does not show in the region of the oil drop on the porcelain, and so one can visually estimate the thin deposit by the contrast differential as seen in this figure. (e) When carbon is to be evaporated on a mica surface to be used as a self-sustaining carbon film, it must have a substantial thickness. Then carbon is deposited until the porcelain is decidedly darkened as indicated in this figure.

chamber. At this point carbon presumably is or soon will be evaporating. One can see the film as it is deposited and judge its thickness by the discoloration of a piece of white porcelain placed at about the same distance from the carbon source as the objects to be filmed. The discoloration can be seen particularly well if a drop of vacuum pump oil has been placed on the surface of the porcelain (Fig. 20(c)). The area covered by the oil will remain white as the carbon film builds up, and by the contrast one can quite accurately assess the amount of discoloration and hence the relative thickness of the film (Figs. 20(d) and (e)).

Once carbon begins to be evaporated it comes rapidly so that the porcelain indicator darkens quite quickly. The operator must keep his eyes on this and be prepared to turn off the current suddenly when it has been suitably darkened. If evaporated carbon is being used to stabilize a plastic film, only a very thin layer will be required, and the current should be shut off as soon as a very light deposit has been built up on the indicator (Fig. 20(d)). In practice, this means shutting off the current just about as fast as one can after the image of the oil droplet is first visible. If carbon is to be the sole supporting film for sections, much thicker deposits are required which the investigator will have to learn by experience. However, the porcelain should be decidedly darkened as in Fig. 20(e).

There is some evidence (Renshaw and Schneidmiller, 1958) that the strongest carbon films are those produced most rapidly at temperatures substantially above the minimum required to evaporate carbon. One can design a holder for carbon rods in such a way that a controlled amount of carbon is evaporated at high temperature, and then the machine automatically shuts itself off. This is done most simply by planning to evaporate a given length of narrow carbon shaft. Tips are made essentially as in Fig. 20(b). A stiff spring drives one carriage into the other, and an adjustable stop is added which can be set to limit the movement of this carriage so that a known short length of the tip can be evaporated, and then the advance is stopped. A predetermined amperage can be applied suddenly by throwing a switch. The carbon tips are heated in the most rapid way possible, and the full complement of carbon is vaporized almost explosively with sparks flying. A disadvantage of the method is that the shower of incandescent carbon chips can damage the film as it is being made.

6.6. Carbon Stabilization of Naked Sections

Carbon can be evaporated directly on naked sections after these have been picked up on screens. The carbon film so formed will stabilize the methacrylate and give it necessary support. A very thin film of carbon suffices, one which barely begins to darken the porcelain indicator (Fig. 20(d)). No other method is capable of producing a more electron-transparent support. There are limitations and inconveniences to the method, however, so that more elegantly prepared supporting films are desirable in many instances.

It is possible that evaporating carbon directly upon methacrylate sections sometimes produces a fine order damage which destroys their usefulness for high resolution work. The damage, when it occurs, probably is from heat coming from the intense bombardment of the methacrylate by the infrared radiation inevitably present as the carbon is evaporated at over 3500° C. This damage is not likely to be detectable at moderate magnifications so the method is quite applicable to survey work and its convenience recommends it. When high resolution work is contemplated, a control is a wise precaution.

Sections supported by carbon evaporated directly upon them may, at first, create a poor impression, for every little hole or tear in the original section remains a hole which is glaringly conspicuous. Even under the most favorable of circumstances, there are ordinarily at least a few small holes in every grid square. But it should not be difficult to find very nicely exhibited low power fields of epithelial tissues. If difficult material has been sectioned, however, particularly if there was much collagen in the specimen, there will be great numbers of large and small holes which will make most low power fields look rather messy.

When a portion of section is not supported on all four sides by grid wires, the evaporated carbon film usually does not offer sufficient support and the section is lost in that square. Thus, if there are many scratches in the sections, most potential squares may be lost and the yield of good areas be small. For this reason the technique is mainly useful with material that has been well sectioned.

6.7. Carbon Stabilization of Plastic Films

Extrathin Parlodion or Formvar films can be stabilized by evaporating a thin coat of carbon upon their surfaces. This gives them strength and prevents the drift or creep that pure plastic films are apt to show. The method is so simple and reliable that it can be recommended as the method of choice for the beginner, and it is apt to remain the best routine way of preparing supporting films.

As thin Parlodion or Formvar films are prepared as will stand up on the supporting screens. To do this, the stock solution of Parlodion or Formvar is diluted as much as one can get away with. Then coated screens are prepared as previously described. Finally, carbon is evaporated on the screens until the film is just visible on the porcelain indicator. Once screens are prepared this way, they can be stored in covered dishes until one is ready to use them. Thus, a considerable number of screens can be made at one time.

The only important disadvantage of this method is that the combined film is slightly thicker than is actually needed. Thus, one does not have quite the contrast and potential resolution that can be obtained using pure carbon or even pure plastic. For very high resolution work the pure carbon films described below are preferable.

6.8. Pure Carbon Films Prepared on a Plastic Substrate

Pure carbon films mounted on grids can be prepared by beginning with grids having Parlodion or Formvar films upon them. Carbon is then evaporated on top of this substrate, and then finally the substrate is dissolved away from the regions of the grid holes leaving just the carbon (Watson, 1955). Obviously, however, the substrate must not be removed completely from between the wire and the carbon film or the latter will float away.

If carbon is to be the sole supporting film, it necessarily must be deposited in a thicker layer than when it is used to reinforce and stabilize a plastic film. One should evaporate sufficient carbon so that the porcelain indicator is decidedly darkened as suggested by Fig. 20(e). The operator will have to gain experience in judging this by trial and error.

If a Parlodion substrate has been used, reagent grade acetone is a satisfactory solvent. Coated screens are slid under the surface of the acetone at a steep angle (at least 45°). They are removed in perhaps 2 or 3 min, but experience and practice will determine just how long they should be left submerged to remove all of the substrate in the vicinity of the grid holes yet not dissolve the plastic between the wire and the carbon film. If Formvar is the substrate, chloroform may be used as a solvent, in which case it will dissolve the Formvar in a matter of minutes, or ethylene dichloride may be used, which will take several hours. When the screens are submerged, it is convenient to place them on a piece of wire mesh (window screening), they then can be picked up without fumbling and delay when it is time to remove them from the solvent.

A virtue of this method is that the carbon films remain firmly adherent to the grid when they are prepared exactly correctly. The disadvantage of the method, however, is that this timing is difficult to standardize, and there is, after all, unavoidably rough treatment as the screens go in and out of the solvent. There are also surface tension forces at work as the solvent finally evaporates. Thus, the method has a number of hazards. One might paraphrase the nursery rhyme as follows: "When this method is good, it is very very good, but when it is bad, it is horrid."

6.9. Stripping Pure Carbon Films

Carbon films evaporated on glass surfaces ordinarily cannot be stripped which has prevented the exploitation of this potential method. As will be seen later, however, specially prepared glass surfaces can be used. However, freshly split mica is a good substrate, and carbon films can be floated from this easily.

Pieces of mica, approximately an inch square, are split just before use. A clean razor blade can be used for this purpose, but be sure not to contaminate the fresh mica surface with oil or grease films that may coat new blades. Enough carbon is evaporated to discolor decidedly the porcelain marker as in Fig. 20(e). Then the film is floated off on a clean water surface by dipping the mica under the water at a low angle. Breathing on the film to "frost" it with condensed moisture droplets seems to help, but ordinarily there will not be difficulty in stripping the film. As one corner of the mica is submerged, the operator should pause and be sure the film is separating. It can, of course, be seen by reflected light. If separation will not start in one place, it probably will in another.

Carbon films that are stripped from mica surfaces do not look particularly smooth to the naked eye as they float upon the water surface. However, on the small scale they are beautifully smooth and uniform.

To mount the carbon films, individual grids are submerged below the water surface while being held by fine pointed forceps. They are slowly lifted out of the water through the floating carbon film and transferred to filter paper for drying.

Unfortunately, carbon films do not adhere well to copper grids. However, it is possible to precoat the grid wires with an adhesive that helps. This author has, from time to time, used an extremely dilute solution of rubber cement in carbon tetrachloride. Or, the adhesive on a short length (about 4 in.) of cellophane tape can be dissolved in an ounce of chloroform. The grids should be dipped in this and quickly dried on filter paper. The adhesive must be so dilute that there is no tendency for it to form films or strands across the grid openings.

Evaporated carbon films apparently do not adhere strongly to a truly clean glass surface, and can be stripped from this base. It does not seem possible to obtain consistently a satisfactory glass surface by chemical cleaning, but extreme heat suffices. An ordinary 1 x 3 in. microscope slide is used. It is heated in a Bunsen flame until the glass actually begins to melt and bend, and the edges become fire-polished. Such a prepared glass surface is used immediately. It has seemed to this writer, however, that it is harder to separate carbon films from this surface than from mica.

Kafig (1958) has proposed an ingenious method of preparing a glass surface so that carbon films can be stripped from it. He advocated first depositing a layer of the dye, basic fuchsin, on the glass and then evaporating carbon on top of this. Subsequently the dye is dissolved and the film can be stripped. The method depends partly upon depositing the correct amount of basic fuchsin in the first place. The stock solution consists of 3 gm of the dye in 150 ml acetone, to which 1 ml nitric acid and ½ ml Aerosol OT 25% detergent is added. A standard microscope slide is dipped in this, drained and dried. After the carbon film has been deposited, the slide is soaked in acetone, acidified by the addition of 5% acetic acid, until the color disappears. The slide is then dried, after which the carbon film can be floated off. (This procedure was not completely described in Kafig's published report.)

Pure carbon films are extremely brittle and will not stand stretching. Thus, grids with carbon films upon them must not be bent unless one is willing to sacrifice the film in some of the squares.

6.10. "Sandwiched" Sections

It has already been stressed that conventionally mounted methacrylate sections may exhibit a fine-order artifact that destroys macromolecular detail (Chapter 4). This results from the melting of the methacrylate during electron bombardment, and consequent surface tension effects. Figure 8 indicates how much of the embedding medium actually disappears during observation, and it is easy to understand how delicate structures at the exposed surface can collapse. Thoughts about this led Watson (1957) to propose that both surfaces of a section be protected by supporting films.

Watson originally used two carbon films for support. However, for convenience, and for maximum transparency, one of the films can be of Parlodion or Formvar. (Do not, though, evaporate carbon through the grid mesh to the back side of a section. No carbon reaches the parts of the section protected by grid wires, and this permits different tensions to be set up on the two sides of the section when it is subsequently examined. The usefulness of the preparation is apt to be destroyed by numerous ripples which then may develop.)

Sandwiched sections preserve at least surface detail elegantly (Figs. 21 and 31(a)), and thus somewhat resemble the preservation achieved with Araldite and Vestopal W embedding. Naturally the second supporting film reduces contrast significantly, and herein lies the principal difficulty with the method. Under almost all circumstances it is necessary to use "stained" material. It is also essential that the supporting films be as truly thin as possible. It may prove desirable to use a high accelerating voltage (100 kv) with relatively thick sections to achieve a good contrast differential.

Roth (1959) has suggested recently that a thin methacrylate film be used as the top layer of the sandwich, for this has the advantage of decomposing somewhat in the electron beam, and

---

FIG. 21. Conventionally fixed kidney mounted sandwiched between two layers of evaporated carbon to avoid the collapse of structures at exposed surfaces which otherwise occurs as methacrylate decomposes and sublimes in the electron beam. (a) A transverse section of a glomerular capillary showing the urinary space (ur.), part of an epithelial cell and its feet (epi.), the basement membrane and fenestrated endothelium (end.). (b) The apical portion of a cell of a proximal tubule showing the basal portion of the brush border. This should be compared with Fig. 2 which had a supporting film on only one side of the section. Although there is a better preservation of detail in the sandwiched section, much contrast is lost, and electron "stains" are apt to be necessary. These sections were floated upon a 0.5% solution of phosphotungstic acid in 50% alcohol before mounting.

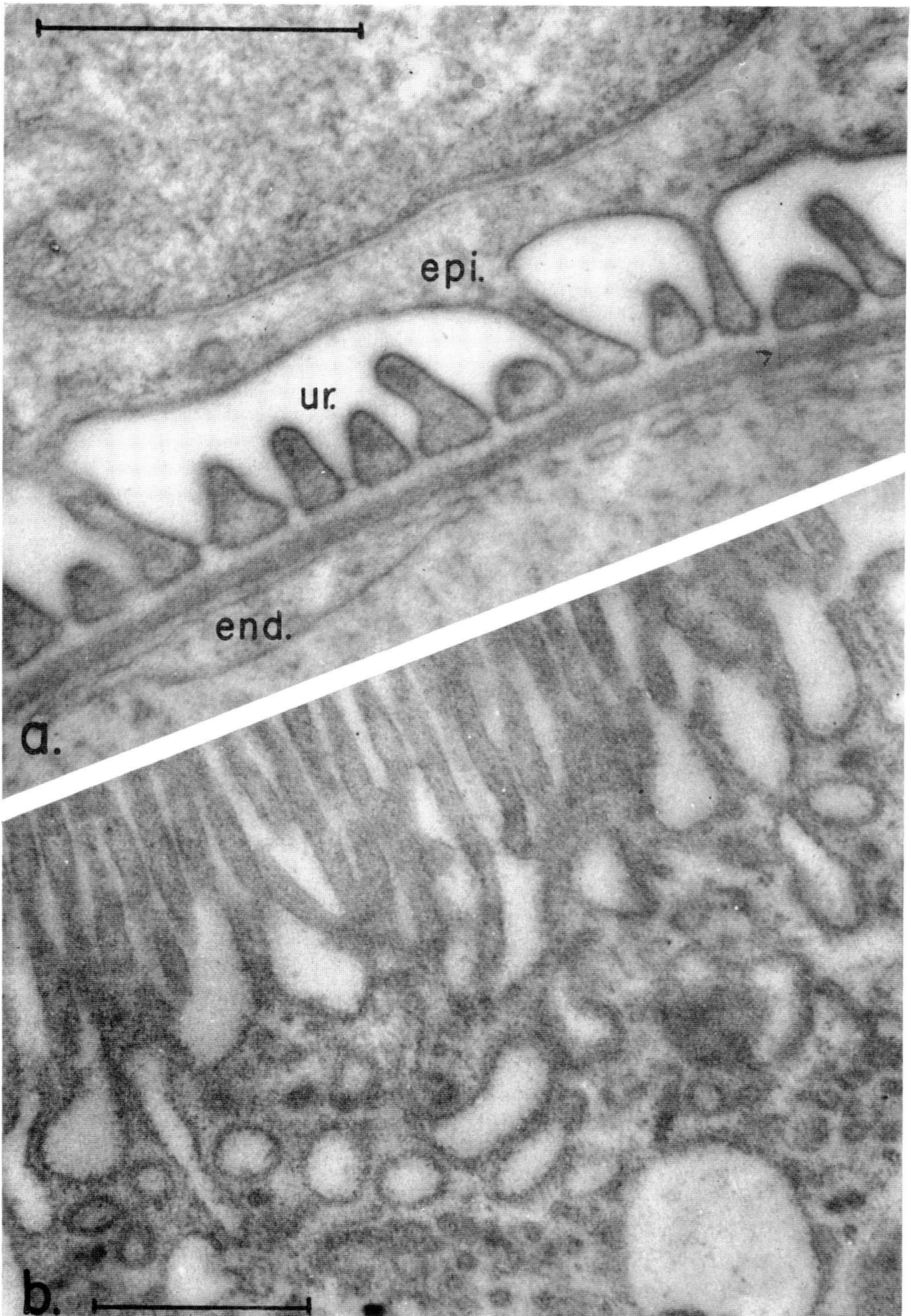
epi.
ur.
end.
a.
b.

thus becomes quite transparent. Yet, apparently, it is adequate to preserve intact the top surface of the section. He indicates that suitably thin methacrylate films can be made if 1–2% prepolymerized methacrylate (Chapter 4.5) is dissolved in methylmethacrylate monomere. A drop of this solution is allowed to spread in an unimpeded fashion on a clean water surface. Grids prepared with carbon substrates, and with sections upon them, are inverted on top of the floating film. They are then picked up from the water surface as described earlier in this chapter, and are ready for examination after drying. It is suggested that the thickness of the added methacrylate film be adjusted so that after electron bombardment the interference colors of the section are the same as they were to begin with. Thus, what originally was a pale gold section might show a dark gold color after being mounted on the substrate. The addition of the methacrylate film would then, perhaps, produce purple, blue, or green colors. However, after the section is bombarded in the electron microscope and some of the methacrylate decomposed, the section should return to the dark gold color.

6.11. Serial Sections

It is not difficult to cut reasonably uniform serial sections. It should be realized, however, that the full series of sections to be examined almost has to be put on a single grid, for hazards multiply in trying to continue a series on several grids. Thus, the block must be trimmed so that the section width is very narrow and the number of sections needed will be included in a ribbon approximately 1/16 in. long.

One cannot use the ordinary supporting grids for serial section work. It is necessary to have grids that have long slots in them rather than square holes. Such patterns are available from the manufacturers of high quality grids listed in Appendix B.

The long axis of the ribbon of sections naturally must be aligned fairly exactly with a grid slot if one is going to be able to follow a structure throughout the length of the slot. Some investigators may be able to pick up satisfactorily oriented sections by hand using a binocular dissecting microscope for the purpose. More elegant methods of orienting serial sections and picking them up have been described, however. These involve some of the principles and equipment of simple micromanipulation.

Gay and Anderson (1954) were perhaps the first to report on a long series of sections. Their ribbons were picked upon a Formvar film supported by a small wire loop. Their grids were placed on a plastic assembly on top of an adjustable condenser in a standard microscope. By racking down the condenser, the grid was kept below the stage level while suitable alignment was achieved with the ribbon on the loop. Then contact was established by lifting the condenser assembly.

A disadvantage of Gay and Anderson's method is that fairly heavy supporting films are required. This can be obviated if floating sections are picked up directly upon grids. Perhaps the simplest way of obtaining a large measure of mechanical control is to attach a simple platform to a low power microscope objective. The platform, which holds the grid, should be just below the focal level of the objective, but close enough to this plane so that an out-of-focus image of the grid slots can be seen. The grid is at first below the surface of the fluid, and the floating ribbon is aligned above it. This may be done freehand or with a simple manipulator. The screen is then lifted to pick up the ribbon simply by racking up the microscope barrel.

Grids with slots do not provide as good a support as those with the more conventional square holes. Also, one cannot afford to take the chance that the supporting film across a particular slot may break, whereas ordinarily if a few squares are lost it is not a matter of consequence. Therefore, supporting films must be somewhat thicker and stronger. Certainly for most work plastic films stabilized with carbon will prove most satisfactory. These will not "creep" as pure plastic is prone to do when it extends long distances, and they are not brittle as is pure carbon which almost certainly will be ruptured when spanning long distances.

6.12. Mounting Screens in the Microscope Specimen Holder

Most electron microscopes have a specimen holder with a cap into which the specimen screen is placed. One soon becomes skillful at "flipping" screens into this. They are balanced on the edges of the cap and then tipped in, turning over as they go.

Screens always should be mounted in the holder with the supporting film and specimen down. It is important to standardize the orientation, for magnification depends upon how close the specimen is to the pole-piece of the objective lens. Calibration will be meaningless if the specimen is not always at the same level. It is preferable to have the specimen down rather than up, to take advantage of as much magnification as possible. Also, in that position, the supporting film is in part protected from the electron beam by the grid wires.

6.13. Dirty Specimens

At times dirt can so contaminate sections that they will be nearly useless. Unless the dirt comes with the specimen screens, it must be blamed on the operator's faulty and careless technique. He has introduced the dirt at one or another step during the sectioning, during the preparation of supporting films, or during the final mounting. Perhaps the most dangerous times are during sectioning and when the sections are being picked up. Dirty forceps tips, needles or spatulas may contaminate the fluid surface upon which the sections are floating. Oil films are the worst offenders. The trough itself or the plastic dish to which sections are transferred may be contaminated in the first place, although there is no real excuse for this if fingers are kept off the inner surfaces. The containers used to hold the acetone and alcohol solution may be a source of contamination. We have found that technical grade acetone may cause trouble and are careful to use only reagent grade now. Also, demineralized water should be employed in making up the solutions; otherwise fine crystalline precipitates may appear as contaminants. The water surfaces upon which supporting films are floated are a dangerous source of contaminants as indicated earlier in this chapter.

Supporting grids may be quite dirty when these are received from the manufacturer. We have had batches that contained much fine copper dust. Other contaminants as well are possible. Batches of dirty screens can be washed, however, and an Erlenmeyer flask of about 200 cc capacity is a convenient receptacle. A quantity of screens may be shaken up in this with a soap or detergent solution. Repeated sudsing will carry away particulate matter. The soap is rinsed away with several changes of demineralized water. Finally, as an aid to drying and also to provide for additional degreasing, they are passed through a reagent grade acetone or absolute alcohol bath and dumped out on large sheets of filter paper to dry.

Sometimes it is desirable as part of the cleaning process to etch screens somewhat to enlarge the holes. A dilute, nitric acid solution can be used as a prelude to the washing procedure outlined above. One should be prepared to stop the action of the acid quickly in case there seems danger of dissolving the screens altogether. This can best be done by adding a large volume of water and decanting.

Literature Cited

Bradley, D. E. (1954). *Brit. J. Appl. Phys.* **5,** 65.

Gay, H., and Anderson, T. E. (1954). *Science* **120,** 1071.

Kafig, E. (1958). *J. Appl. Phys.* **29,** 1624.

Renshaw, T. A., and Schneidmiller, R. F. (1958). *J. Appl. Phys.* **29,** 1624.

Roth, L. E. (1959). Proc. Elect. Micro. Soc. Amer., *J. Appl. Phys.* **30.**

Sjöstrand, F. S. (1957). *In* "Electron Microscopy: Proceedings of the Stockholm Conference, September 1956" (F. S. Sjöstrand and J. Rhodin, eds.), p. 120 Academic Press, New York.

Watson, M. L. (1955). *J. Biophys. Biochem. Cytol.* **1,** 183.

Watson, M. L. (1957). *J. Biophys. Biochem. Cytol.* **3,** 1017.

7. Microscopy

7.1. Introductory Remarks

The biological investigator working with ultrathin sections is involved in a never ceasing struggle to improve contrast. For him, contrast is an intrinsic difficulty since his material inevitably consists of substances made up of light atoms having approximately the same density as the embedding media and supporting films. It would be fair to say that any given specimen has a certain potential contrast which we can only partially realize by the most skillful microscopy and modern photography.

To a large extent contrast depends inversely upon the amount of "background fog" that the photographic plate records. Fogging results from scattered electrons, one way or another reaching the photographic emulsion. These electrons travel at varying speeds, and therefore are not focusable. They do not particularly traverse the refined optical pathway that we are interested in, but many spray the plate randomly after having bounced off the sidewalls of the microscope column. It is for this reason that very important steps can be taken to minimize the number of scattered electrons which finally reach the viewing screen and photographic emulsion. If apertures of suitably small size are placed at critical levels along the microscope column, the useful beam can go through these tiny holes while scattered electrons, even though they may be nearby, will hit the aperture plate and effectively be removed.

The original post-war microscopes were not apertured at all in the modern sense. Such microscopes are perfectly useful today if used solely with specimens of great intrinsic contrast. These include particulate materials and surface replicas which have been shadowed with heavy metals. The first type of preparation is commonly used by bacteriologists and virologists, the second type, by metallurgists. Unapertured microscopes are also used for the direct observation of dense particulate matter as in dealing with pigments, smokes, etc. Investigators in these fields have a much easier time with their microscopy than those of us who are interested in tissue work with ultrathin sections.

The widespread use of objective apertures pretty much coincided with the development of ultrathin sectioning (Fig. 22). It was the RCA EMU-2C microscope (which appeared in 1950) which first had this as part of its regular equipment. The aperture was mounted between the lens elements. Mechanically this presented very serious problems of optical alignment, for the aperture, necessarily, had to be exactly in the path of the main electron beam. Originally one had to remove the pole-piece every time an adjustment was to be made. It might take a half day or more of extremely vexing work to achieve the initial alignment of a new aperture disk. In the meanwhile the chances were that the edge of the aperture had become seriously contaminated, introducing other problems which will be discussed later. This most annoying mechanical difficulty was allayed only partly by the realization that a slightly loose aperture disk could be bounced around, and so centered while the microscope was operating, if one used a rubber mallet to wallop the column. Some investigators become quite skilled at this ludicrous but useful maneuver.

Several independent efforts were made to devise a refined way of moving and aligning objective aperture disks while the microscope was in operation. Most American microscopes at this time were RCA EMU-2 instruments, and in 1955 the Canal

---

FIG. 22. The same specimen of guinea pig marrow was micrographed before and after a 50 μ aperture was put in the back focal plane of the objective lens. (b) Without an aperture, was printed from an adequately exposed but very flat negative on the most contrasty grade of Kodabromide enlarging paper. The image is very flat and weak. (a) With the aperture, produced such a contrasty image on the negative that a softer enlarging paper had to be used. The central cell in these figures is a neutrophilic myelocyte (neut.). At the bottom of the figures from left to right are the corner of a neutrophilic metamyelocyte, a normoblast (norm.), and a myeloblast (m.'bl.). At the top of the figures is the margin of an erythrocyte (r.b.c.).

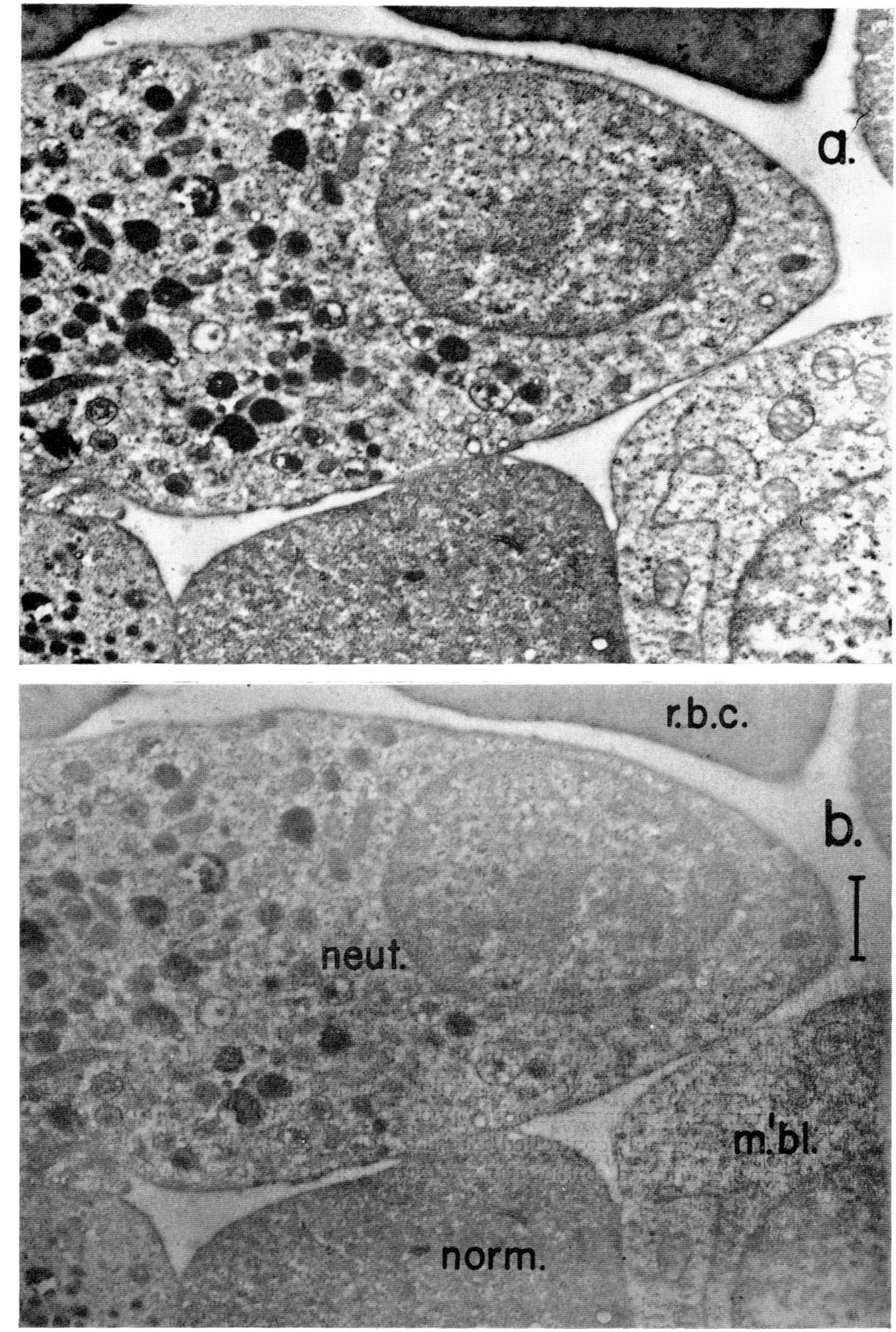
a.
r.b.c.
b.
neut.
m.bl.
norm.

Industrial Corp. (Canalco), Bethesda, Md., introduced commercially such a fitting for these microscopes. This was, indeed, a most important practical development for us. The acceptance of this externally adjustable aperture holder was such that nearly every electron microscope designed since that time incorporates a mechanism to allow external aperture alignment.

The use of very small objective apertures introduced a new difficulty, for when their lips become asymmetrically contaminated, asymmetric electrostatic forces appear. These, superimposed on the magnetic lens, produce astigmatism which quickly deteriorates an image to the point of uselessness. It had been recognized previously that the all-important objective pole-piece could be "compensated" quite exactly by moving soft iron screws between the lens elements in such a fashion that intrinsic asymmetries in the magnetic flux of the pole-piece were exactly counterbalanced (first RCA model equipped this way as a standard fitting was the EMU-2D). With sufficient care, practically all traces of intrinsic astigmatism could be removed by this method. To this day, the investigator whose specimens do not require the use of an aperture can use his compensated pole-piece for months or even years without problems of adjustment, but those of us who must use an aperture are confronted daily with the possibility of astigmatism, and the necessity of correcting for it as it appears.

Of course, one can replace apertures with the expectation (often in vain) of eliminating astigmatism. The very nature of the alignment procedure endangers the aperture, so that by the time a new one is aligned it may well be contaminated and be a source of new astigmatism. Thus, until 1956, we were working precariously at best and often used the microscope knowing that some astigmatism was present, but felt that we would not necessarily improve things by attempting corrective measures. Then the Canal Industrial Corporation added to their externally centerable aperture holder a group of six radially arranged, nonmagnetic pins, opposing pairs of which could be given a variably controlled and stable electrostatic charge. This can be adjusted so that its asymmetric effect is set to counterbalance the astigmatism of the lens system. Thus, the lens can be compensated while the instrument is in operation simply by turning two knobs, one changing the azimuth of superimposed charge, and the other controlling its magnitude. From the practical point of view this has been a very important development for now one can correct astigmatism as it appears and with little effort. There is now no excuse for compromising with quality. Modern microscopes include this feature or its magnetic equivalent.

Although the objective lens is the very heart of any microscope and deserves primary attention, the optical performance of modern electron microscopes has been improved in yet other ways. Although a properly placed, suitably small objective aperture does much to remove scattered electrons, some get through even this tiny hole and, indeed, with the main beam itself. It became apparent that contrast could be improved further by placing additional apertures in the projector pole-piece. These apertures fortunately need not be very small to be nonetheless effective. Simple mechanical alignment is usually enough, and contamination is not a serious problem with them. In general, modern microscopes come reasonably well equipped with such apertures, and the investigator accepts them as part of the basic instrument without giving them further thought.

Small apertures are now also used in the condenser lens, partly to increase the contrast, but here they have at least one additional important function. "Space charge" effects occur if electrons are crowded too closely together. This can easily happen as an intense beam passes through a focal point. If this should occur in the objective lens near beam "cross-over" it destroys resolution. This is avoided by limiting the possible beam intensity by using a condenser aperture of suitable size. Some modern microscopes also refine the beam considerably before it reaches the specimen by using more than one condenser lens.

In the subsequent portions of the chapter no attempt will be made to consider how particular microscopes should be handled or maintained. These details naturally vary much from one manufacturer's product to another, and this sort of instruction and maintenance is available from their handbooks and service departments. On the other hand, as these introductory remarks have indicated, those of us working with tissues have special problems which do need understanding and detailed consideration. These are in addition to certain of the general problems of maintenance which every microscopist must master in a practical way. To operate a microscope successfully and run a laboratory, the investigator must be able to take care of the microscope column himself. He cannot rely on the manufacturer's service department to do this for it requires day to day watchfulness, and minor adjustment. The operator must know the warning signs of impending difficulty. Usually he will be able to correct it himself and at least save time, but also he must know when he needs help and when to call a service man, engineer, or physicist. Thus, a portion of this chapter is devoted to commonplace problems of maintenance without reference to particular types of instruments.

7.2. Objective Aperture Position and Size

There are two places where a small aperture can be placed in an objective pole-piece to screen out effectively most of the scattered electrons. It may either be between the two lens elements of the pole-piece or be in the back focal plane. In the between-the-lens position the cross-sectional diameter of the main beam is particularly small, and apertures of 20–50 μ can and should be used. The smaller the aperture, the better the final contrast, but one pays for this in that the contamination rate is increased and astigmatism becomes a very serious problem. To avoid asymmetrical contamination, the intense crossover beam must never be allowed to touch the rim of the aperture. Apertures approximately 35 μ in diameter are a reasonable compromise, being large enough to accommodate the crossover beam with a small margin of safety, yet providing good contrast.

Between-the-lens apertures are fairly close to the specimen.

They are quite vulnerable to materials which may evaporate from this, most particularly the decomposition products of methacrylate. Thus, they tend to contaminate unduly rapidly in this position. Apertures in the back focal plane of the lens are much further removed from the specimen and their contamination rate is slower. The cross-over beam is wider here, however, than in the between-the-lens position, so that the apertures must be correspondingly larger. Apertures of 50 μ in the back focal plane correspond approximately to 35-μ apertures in the other position (Fig. 22).

If an aperture is used in the back focal plane, it is important that it really be near the level of this plane. If it is either too close or too far from the lens it will restrict the optical pathway so that the edges of the aperture will be seen in silhouette in the final image. Even more importantly, however, its rim will be bombarded unnecessarily heavily by a nearly cross-over beam, and it will be impossible to avoid astigmatism for long.

The position of the back focal plane depends upon the focal length of the lens. Unlike glass lenses which have fixed focal lengths, magnetic lenses, of course, are varied at will simply by altering the current flow. It is necessary that the aperture be in the back focal plane when the lens system is operating in a manner to focus a specimen. In getting ready to find this position, a specimen first is focused at a fairly high magnification, and the position of all of the lens control knobs noted. The height of the aperture is adjusted to correspond with this particular focal length.

To find the back focal plane, one takes advantage of the fact that the aperture edge can be seen in silhouette except when the aperture is in the proper position. One starts with an old aperture in the holder, and deliberately moves its rim into the beam so that it can be seen as a somewhat out-of-focus silhouette on the viewing screen. The operator will not know at first whether the aperture is too close to, or too far from the lens. He then uses the coarsest focus control and goes up or down the scale. The image of the aperture lip probably will tend to move off the field of view and will have to be kept in view with its alignment screws. At some level of lens current the rim of the aperture will completely disappear from view and, as the controls are turned further, it will reappear again but with reversed symmetry. Thus, there is a "cross-over" effect which appears when the aperture passes the back focal plane. This will be the clue as to whether the aperture is too close to the lens or too far away from it. If the lens is operating as an unduly weak lens at low current levels, the gap between the aperture and the lens is too long, and the holder must be elevated. Conversely, if the lens current is above normal so that the lens is too strong, the aperture is too near the lens, and the gap must be lengthened. Successive adjustments will have to be made until finally the aperture is close to the back focal plane when a specimen is also in focus. Ideally, the aperture height should be adjusted so that it is in the back focal plane when the instrument is in focus for a fairly high magnification.

Fortunately, once the correct aperture position has been determined, it will not need readjustment unless the specimen distance from the pole-piece is altered. This means, however, if more than one specimen holder is used, that all be positioned so that the specimen plane is identical. It may preclude the use of stereoholders.

7.3. Objective Aperture Alignment

It has already been indicated that when apertures become asymmetrically contaminated, astigmatism appears which ruins good resolution. Inevitably, apertures have to be replaced fairly often, which means that microscopists must become adept at their alignment. The great hazard in this is that contamination will occur as the aperture is being aligned if a high intensity beam is allowed to bombard the rim of the aperture during this process. It is necessary to work with a greatly spread beam, and probably also with a beam current substantially below saturation.

If a between-the-lens aperture is being used, it is not possible to obtain a focused image of its edge. However, at low magnification a circle of illumination can be seen having a fuzzy limit. The center of this should be made to coincide with the center of rotation of the objective lens which then represents alignment.

Naturally when a between-the-lens aperture is first installed it may be so far out of alignment that no illumination can reach the stage. It is wise to have anticipated this, and thus be able to set up the operation of the microscope in a manner that avoids damage to the misaligned aperture. For this, one also needs assurance that the screen could be weakly illuminated if the aperture were only centered. Therefore, before starting to replace an aperture, the beam should be spread widely, and perhaps the beam current reduced, and these settings recorded so they can be duplicated. If a specimen is in the microscope, be sure it is not moved subsequently, or a grid wire may block illumination. Return to these original conditions before starting to search for the aperture. Have confidence in them, and do not start making other adjustments. It is only after a reasonable degree of alignment is achieved that the beam may be brought close to cross-over.

A rear focal plane aperture can be aligned more readily and exactly than a between-the-lens aperture since it is possible to get a quite satisfactory image of its rim, and the position of the beam within it. To do this, one resorts to an optical trick. The microscope is set up as for selected area diffraction*. A cross-over beam (but with very low, unbiased beam current) is passed through any sort of material which will diffract electrons. The beam is focused so that it appears as a tiny bright spot. The neighboring areas are illuminated diffusely by a cloud of diffracted electrons. The rim of the aperture is silhouetted against this fairly sharply (Figs. 23(a)–(c)) having an apparent diameter of $\frac{1}{8}$–$\frac{1}{4}$ in. Naturally, when a new aperture is first introduced, its disk may hide the bright beam. One first seeks the area of

* The RCA EMU-2 microscopes are not equipped for selected area diffraction except for the final model (2E). But the Canal Industrial Corporation markets a "Centering Guide and Diffraction Device" as an accessory for these microscopes which creates the desired optical conditions for aperture alignment.

diffuse illumination, and then, by manipulating the controls, an off-center aperture (Fig. 23(c)) can be exactly aligned (Fig. 23(b)), so that the beam spot is in the center of the illuminated circle.

This method of centering is accomplished easily if anticipatory precautions are followed. Before beginning aperture replacement a suitably diffracting object should be set up. A thick section will do, or perhaps a fold in a section or a place where there is contaminating dirt. If necessary, one can prepare special screens with carbon or magnesium oxide smoke deposits. Obtain the focused image that will be used later for centering the new aperture and note the beam current at saturation and also the condenser setting at cross-over. Then reduce the beam current until the background illumination is faint, and only the cross-over beam remains as a fairly bright spot (Fig. 23(a)), but its intensity should have been so reduced that it would be relatively safe to touch the rim of a new aperture with it. This is the image that will finally be used for centering the new aperture, and the beam current necessary to obtain it should be noted. Next, spread out the beam as far as possible with the condenser control and increase the beam current until a faint diffuse illumination is all that is visible. Note

---

FIG. 23. When the lens system of an electron microscope is used as for selected area diffraction, and a very small beam is allowed to pass through a diffracting material, one can see the position of the cross-over beam in relation to the silhouetted rim of an objective aperture. The image one actually sees is about the size of (a). The other figures are greatly enlarged. The small beam is aligned with the center of the aperture in (a) and (b). In (c) it is decidedly off-center, bombarding the aperture rim, and requires adjustment.

In order to set up the optimal conditions which make possible alignment of the aperture, the beam must be as small as possible. A cross-over beam is used, but with very low beam currents. The beam must be focused using the diffraction focus control so that it becomes as small as possible. When the beam is slightly out of focus, or is not at the cross-over position, one may see images such as those of (f) in which a distorted image of a hole in the specimen screen is apparent with a torn portion of a section within it. Obviously beam conditions in the upper right hand corner were approaching those necessary for alignment. All that was required was further focusing. In (d) and (e) the cross-over spot was too large and could not be focused to yield the necessary thin pencil of electrons. ((d) is focused, (e) is off-focus). Under these circumstances success may be achieved by hiding part of the cross-over beam behind a specimen wire, and thus reducing its width.

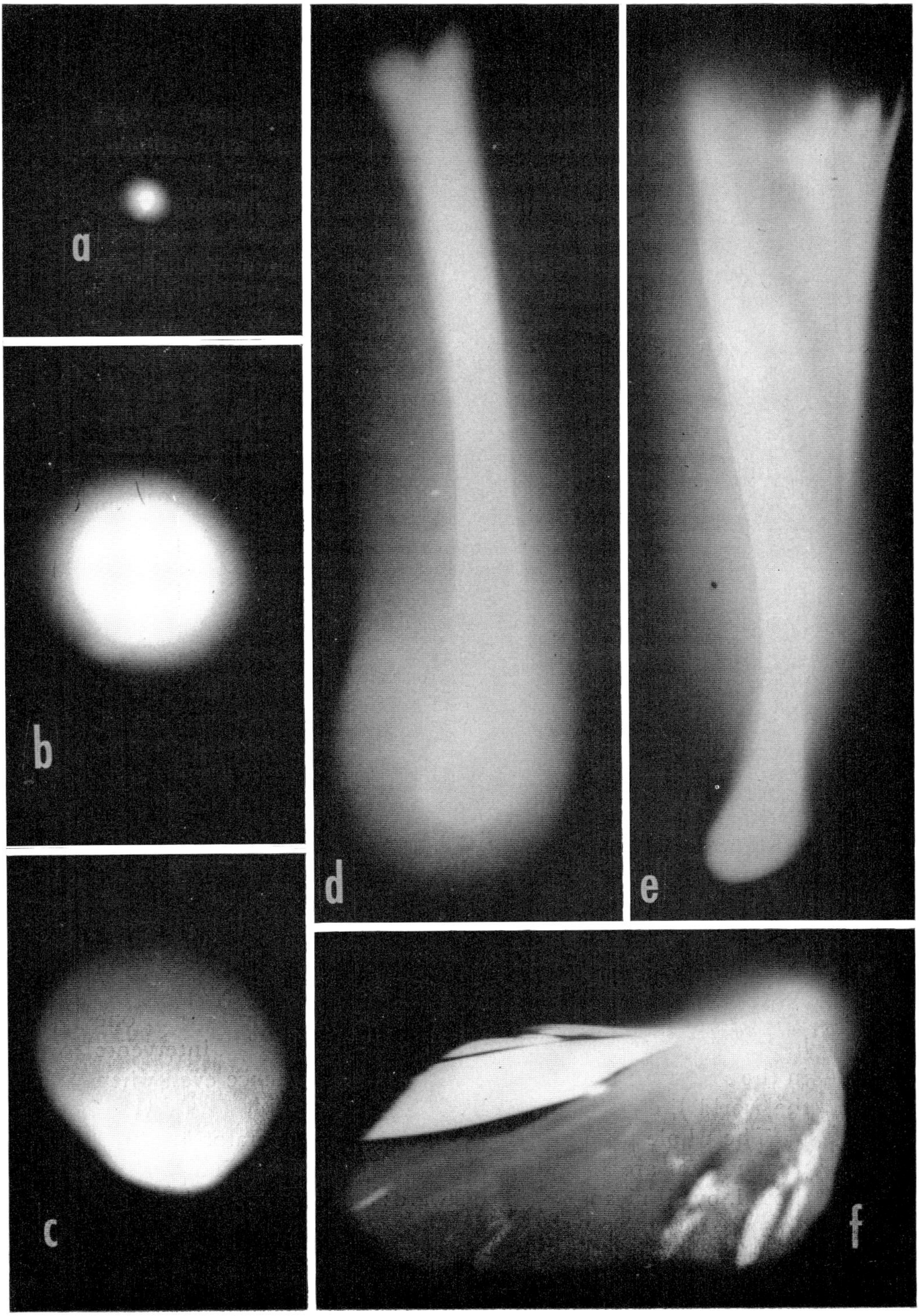
a
b
c
d
e
f

this beam current, and return to this when the search is started for the new aperture. This low order diffuse bombardment should not harm the new aperture.

When the new aperture is first put in, the microscope is returned to conditions which are known to be safe and also capable of providing the images necessary for alignment. One first finds the illuminated image of the aperture rim. Then one begins to condense the beam towards the cross-over position, reducing the beam current as this is done to keep the aperture barely visible. As the cross section of the beam is reduced the aperture is moved correspondingly to keep it in view. After several successive adjustments a cross-over beam will be achieved within the boundaries of the aperture, and final alignment is close (Figs. 23(b) and (c)). Only after alignment can the beam current safely be increased to the bias position.

The experienced worker and the beginner will alike be frustrated at times in trying to get any illumination after installing a new aperture. It may be so far out of alignment that nothing appears on the fluorescent screen even when the beam is maximally spread out. The operator may attempt moving the aperture about with its controls with the hope of picking up illumination. He may be successful but is apt to become completely lost, and there will be no alternative but to open up the microscope column and start again with the aperture centered as best one can by eye. Under these circumstances, there is a temptation to make other adjustments in the microscope, to move the stage or to move the gun translation. In general, these temptations should be resisted, for one is apt to become hopelessly lost. Aperture alignment should begin with a system known to be illuminable from top to bottom excepting only that the aperture disk may be interposed. It is only when one has the assurance that nothing but the aperture disk intervenes in the system that one can proceed with confidence even when the world literally looks dark.

The whole procedure depends upon the operator knowing what images are required and how to get them. The novice should, by all means, work backwards and practice with an old aperture to explore the variables. Figures 23(d)–(f) indicate some of the image patterns he may see. To obtain a satisfactory image of the cross-over beam within the outline of the aperture, the following conditions must be fulfilled. One must begin with a satisfactorily diffracting object. The beam current should be substantially reduced (otherwise the central illumination is so brilliant that little else can be seen). The cross-over beam must be focused to a minimal size by the selected area diffraction control knob (otherwise distorted out-of-focus images of the whole field will be visible as in Fig. 20(f)). To avoid images such as those of Figs. 22(d) and (e) it is sometimes necessary to hide part of the cross-over beam behind a specimen wire, for otherwise it may not prove possible to obtain a small enough spot to avoid the interfering "tails."

When microscopes are used as service instruments for a variety of tasks, it may be necessary to take the aperture out quite frequently. All who use the microscope must be made aware of the necessity of protecting the aperture from intense beam bombardment. Also, one should not attempt voltage alignment of the objective lens when a good aperture is in place, for when the accelerating voltage is reduced, the intense beam spirals away from the original optical axis in a manner which can be expected to intersect the lip of the aperture. Thus, the aperture quite probably will become asymmetrically contaminated.

7.4. Aperture Cleaning and Manufacture

Whether or not an aperture can be cleaned and reused depends partly upon what it is made of. Platinum apertures have become most popular because of the inertness of this metal. The contaminating deposits which form during electron bombardment, however, are almost as insoluble as platinum itself, and there is no really satisfactory way of getting rid of them. Most simply, platinum apertures can be heated in the oxidizing portion of a Bunsen burner flame. Platinum will not vaporize at this temperature, but some contaminants will, and relatively clean apertures result. However, ashes and perhaps amalgams remain which accumulate and sooner or later destroy the usefulness of the aperture. It has been claimed that it is safer and better not to flame the aperture directly, but to protect it by having it in a platinum crucible or over platinum foil. An old aperture can be used in this way to protect a good one.

Platinum apertures can also be heated *in vacuo* in a tungsten boat to a temperature just below vaporization. This can be judged by the incandescent color of the aperture. To learn when to stop heating the aperture one should first deliberately vaporize an old aperture or two while observing the color changes before this occurs.

A variety of chemical cleaning procedures have been proposed. Fused potassium hydroxide works fairly well. Fuming sulfuric–chromic acid mixtures are used. The RCA laboratories recommend the following somewhat elaborate procedure. A platinum or quartz crucible is required. Pure potassium bisulphate is melted in this, and the aperture soaked for at least 5 min. The aperture is then rinsed at least four times in boiling concentrated sulfuric acid. It is finally washed several times with distilled water. The Siemens Company recommends a nearly identical procedure.

Unfortunately not one of these methods will reliably remove all deposits. Even when an aperture looks clean it may produce astigmatism when it is reintroduced into the electron microscope. Thus, there has been more than a little dissatisfaction with platinum, and other metals have been used.

Molybdenum apertures recently have become available commercially. This material has a very high melting point (2600° C, 850° more than platinum). Of course if it were to be heated in air it would oxidize and be ruined long before this temperature was reached, but it can be heated nearly to this great temperature *in vacuo,* and all significant contaminants are vaporized. Perhaps the simplest way of doing this is to use a narrow strip of 2 mil molybdenum as a boat to hold the apertures to be heated. It may be creased down the middle as a gutter, and the apertures lined up along this. The boat is mounted in the vacuum chamber between two electrode posts. After a satisfactory vacuum has been obtained, current is passed through the boat, heating it to a white hot temperature, a little below what is expected to vaporize the molybdenum. It is left at this temperature for perhaps 5 min, and then the current increased by about 5 amp. After another 5 min interval, once more there would be a 5-amp increment. This continues until a segment of the ribbon vaporizes, and the boat breaks. The apertures by this time are lightly welded to the boat so that they do not fly all over the vacuum chamber. After cooling and returning to atmospheric pressure the apertures can reliably be regarded as absolutely clean. Other investigators use a molybdenum or a tungsten boat of standardized dimensions, and heat this with a predetermined amount of current as high as they dare go. However, since the temperature is kept well below that which vaporizes molybdenum, much longer periods of cooking are recommended, periods ranging up to several hours.

Molybdenum may well be the preferred material for making apertures in the future. There have been some complaints with the first ones that they were not bored smoothly. The rough surfaces contaminated very quickly, much more quickly than platinum apertures with a smooth bore. Already it would seem that some manufacturers have overcome this difficulty.

Another approach to the problem of having an uncontaminated aperture is to make them so easily and cheaply that they can be discarded as soon as they start causing trouble and be replaced with a brand new one. If copper is used, this can be done.

Making copper apertures may seem like a formidable task, but actually it is quite simple, and a little practice makes one nearly perfect. A ribbon of 2 mil copper should be cut so that it is slightly narrower than the cap that will hold the aperture. A carefully sharpened needle mounted in a pin vise is used to bore a tiny hole. The needle is lightly sharpened before each use with a very fine abrasive stone. It is cleaned of abrasive particles by punching it through a piece of paper. The tip of the needle is centered on the ribbon, the proper distance from one end, using a binocular dissecting microscope (Fig. 24(a)). The positioning need only be approximate, and is easy to accomplish. The hole is bored by holding the pin vise vertically and turning it with light pressure. A piece of aluminum makes a good backing material. The hole should be of such a size that light is barely visible through it. The tip of the copper strip is then etched in a bath of 40–50% nitric acid. This enlarges and smooths the hole, removes the burr, and determines the final size. Etching must not be excessive, however, or the hole will cease to be round. The original bored hole should be of such a size that no more than 15 or 20 sec are required to complete etching the aperture and bring it to the desired final size. The etching process is stopped by dipping in demineralized water, blotting, and finally drying. At first one will wish to measure the holes produced with a conventional microscope and a stage micrometer. It is not difficult with practice to make apertures within 10 or even 5μ of the desired diameter. They can be made so quickly and easily that if some are the wrong size or asymmetrical they can be rejected at once and another one made.

When a good vertical, smooth, round hole of proper size is obtained, this end of the ribbon is held in a pair of jewelers' forceps, and the aperture area clipped off (Fig. 24(b)). The small square is mounted in the aperture cap with the tapered bevel facing away from the electron source to protect the sidewalls as much as possible from beam contamination.* In an hour's time

* When apertures are bored with a tapered hole, it is important that they be oriented with the taper away from the electron source. The electron beam, therefore, passes through the narrowest diameter with a relatively sharp lip first, and there is no direct bombardment of the tapering walls by the beam. Thus contamination is minimized; whereas if the aperture were mounted with the taper up, the whole length of the bore wall would receive direct bombardment and contamination would be maximal.

FIG. 24. (a) A very sharp needle mounted in a pin vise is being used to bore an aperture hole in a strip of 2 mil copper. An aperture cap is in the lower right to indicate size. Notice that the strip of copper is narrower than the diameter of the aperture holder so that when the aperture area is finally cut off as a square in (b) it will fit in the cap without clipping its corners. A piece of aluminum is used as backing during the drilling operation. Drilling is followed by an acid etch and thorough washing before cutting the square.

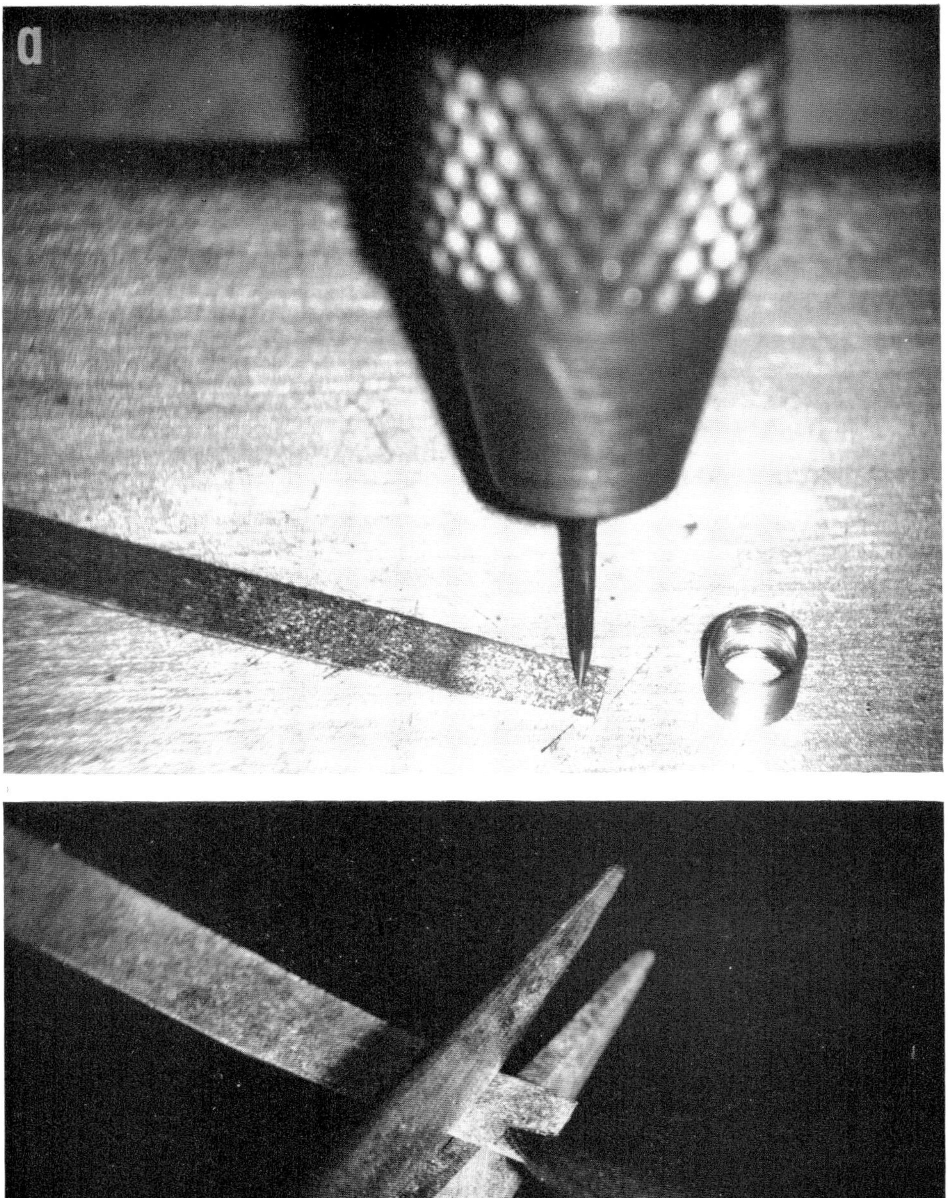
a
b

one can make quite a number of good apertures in this way which probably do not spoil even when kept for long periods of time in gelatin capsules, although the possibility of contaminating oxide layers has to be considered. This writer, who has been using copper apertures of this sort for several years, prefers to make them as needed so that only bright apertures are put into the microscope.

7.5. Microscope Maintenance

Certain aspects of microscope maintenance must be assumed as responsibilities by any critical investigator. He will not be content or happy with what service personnel can provide, even if such help is readily available. Quality microscopy is impossible without keeping a constant check on the performance of the instrument. Good maintenance becomes an essential skill of a good microscopist.

The previous part of this chapter has already introduced the most important problems of maintenance, achieving a well compensated objective lens, with a suitably small clean aperture properly aligned in the objective pole-piece. In addition, the electron microscopist will be faced with frequent filament replacement and its alignment. The next most frequent task will be tracking down bits of dirt and lint which have lodged somewhere in the column and which cause trouble because of electrostatic field effects. This will mean pulling apart various parts of the column, and it becomes essential to know how to realign all parts of the column. Finally, circuit instability may appear. A microscopist must learn the warning signs and he probably will wish to do some of this trouble-shooting himself. Aspects of these problems are considered in the next sections.

7.6 Alignment, General Remarks

It is not practical to discuss details of alignment in terms of particular instruments. Manuals published by the different manufacturers and their service representatives are in a better position to instruct in these mechanical operations. Principles are the same in all instruments, however, and the effects of misalignment nearly identical.

The newcomer to electron microscopy almost certainly will feel overwhelmed by the complexity of alignment procedures if he feels obliged to master these all at once. He can rest assured that in practice alignment is not a serious problem. One has to realign the whole column only when it has been torn down. As a general rule, only one or two alignment procedures are necessary to put the microscope back in operation after a limited maintenance operation on one or another part of the column. Thus, when a filament is replaced, or the objective pole-piece removed to clean the aperture, the alignment of most parts of the column is not affected at all, and realignment operations are limited to the gun in the first case and the aperture holder in the second. It may be many months before the whole column need be disassembled and require complete realignment. From a practical point of view alignment consists of many little problems rather than one big one.

The newcomer to electron microscopy should ask two questions. First, why is alignment necessary, and second how exact must it be in practice for reasonable performance?

There are several reasons for column alignment, some practical, some theoretical. An important theoretical reason is that both magnetic and electrostatic lenses are fundamentally spherical lenses. Inevitably such lenses distort. The distortion becomes increasingly severe as one moves away from the central axis. There is no practical probability that this spherical aberration can be corrected in electron microscopes as can be done when complex glass lenses are used in light microscopes. In practice, however, one need not be very concerned about spherical aberration unless one is trying to make accurate measurements, for it is not particularly noticeable except at very low magnifications.

A very important theoretical and practical reason for (voltage) alignment is that it minimizes the effects of circuit instability. Electrical stability is a relative thing. A slight ripple of the accelerating voltage produces electrons moving at different speeds. This corresponds to using light of different wavelengths. It leads to the electron equivalent of chromatic aberration. The aberrant wavelengths focus at different levels and tend to blur the image thus destroying resolution. Furthermore, electrons going through a magnetic or electrostatic lens move in spiral paths, and not as light waves do through glass lenses. Thus, if there is any change in the accelerating voltage, or if lens currents change, the image turns about the optical axis. Naturally such a movement, even though slight, would tend to destroy resolution, and the effect increases rapidly as one moves away from the optical axis. If there is a reasonable degree of electrical stability in the microscope circuits, this effect is not important at low and moderate magnifications, but it becomes increasingly serious as one strives for high resolution.

A fairly good alignment is necessary just as a matter of convenience. In the aligned microscope one can change magnifications, vary illumination, make large focus changes, and so forth, all without having to readjust the source of illumination much, or recenter the image. In this practical sense, the alignment of the condenser aperture and the intermediate lens are most critical. Fortunately their alignment is quite simple.

Full column alignment involves all of the following elements: (1) the electron source (filament tip); (2) the anode; (3) the condenser lens (sometimes consisting of two independent lenses); (4) the objective lens; (5) the intermediate lens; (6) the projector lens; (7) the viewing screen and photographic plate. Furthermore, it has already been stressed that the histologist will have a movable aperture in the objective lens, and he may very well have one in the condenser lens system. Each of these parts must be aligned with its immediate neighbors. In a very few simplified electron microscopes one or two of these parts may be missing. In some other instruments certain of these components may be coupled together. Thus, usually, the projector lens and the viewing screen are coupled. In the RCA EMU-2 microscopes, the intermediate lens is coupled with the projector lens. It is usual to have most apertures fixed in an instrument excepting those in the objective lens and perhaps the condenser lens. Coupled elements actually represent compromises with true alignment. The next sections consider the particular problems of the different movable elements.

7.7 Objective Alignment

The objective lens is aligned with the projector lens. Two types of alignment actually are possible. The one which originally was recommended is termed "current alignment." Because electrons spiral through a magnetic lens, if the focusing current is varied, the image is seen to spiral about an optical axis. For current alignment, the objective lens was moved about on top of the projector lens until this optical axis coincided with the center of the viewing screen. Unfortunately this axis will not necessarily correspond very well with the axis observed when electrons of differing speeds pass through the lens. The latter is termed the axis of "voltage alignment." For reasons discussed in the previous section, voltage alignment is of considerable importance when one is striving for great resolution. Thus, practically all microscopists today advocate voltage alignment rather than current alignment. The price one pays is a slight inconvenience, for the image can be expected to move as one focuses, since one is not apt to be looking at the center of this rotation.

Voltage alignment is achieved by deliberately varying the accelerating voltage and finding the center of image rotation. The objective lens is then moved upon the projector lens until this axis is centered in the viewing screen. In some of the older microscopes it is necessary to turn off the high voltage switch and observe the expanding image as the high voltage condensers discharge to see the axis of rotation. More modern instruments provide a special switch which makes a moderate, controlled alteration in the accelerating voltage which is a great convenience.

If the intermediate lens is not coupled to the projector lens and therefore must be aligned separately, it must not be functional when voltage alignment is done. In one way or another it must be disconnected. In the RCA EMU-3 and EML microscopes, one position of the magnification control cuts out this lens (taps 7 and 8, respectively). Voltage alignment is done at these positions. In the RCA EMU-2 microscopes the coupling of the intermediate and projector lenses means that only a compromise voltage alignment is possible. It is important to do this with the microscope set at as high a magnification as one expects to use commonly.

Voltage alignment should not be done with a good objective aperture in place. The aperture will surely become asymmetrically contaminated by the beam sweeping across its rim when the accelerating voltage is changed.

Once objective alignment has been achieved, it will not change except as the electrical circuits change. But there will be gradual changes as reference batteries deteriorate, and there may be sudden changes when tubes or other defective parts are replaced. Thus, when any work has been done on the fundamental regulating circuits, the high voltage circuits, or the objective lens circuit, the objective alignment should be checked. It is also wise and convenient to check the alignment every time an objective aperture is removed for cleaning or replacement. Taking out the objective pole-piece and replacing it will make small changes in the optical axis, but if the pole-piece is returned as nearly as possible to its original position these should be so minor that they can be ignored and, indeed, have to be if the pole-piece is apertured. Microscopists whose work demands that they strive for maximum resolution may feel that more frequent alignment checks are necessary, but most work with sections will not be handicapped by a minor objective lens misalignment.

After the objective and projector lenses have been aligned, the intermediate lens (if not coupled to the projector lens) is added functionally to the system. In most microscopes this is used as a reducing lens to extend the lower ranges of magnification. For this purpose one wants alignment for operational convenience. One turns the magnification control in its lower ranges and observes the center of rotation of the image. Once again this indicates the optical axis of the lens, and the lens is moved about until this coincides with the center of the viewing screen (current alignment).

In some microscopes the intermediate lens is used additively at high magnifications. If one regularly uses this end of the magnification scale, it may be desirable to align the intermediate lens for this, rather than for low magnifications, since it is not likely that one alignment will do very well for both.

7.8. Alignment of the Illuminating System

The previous section dealt with the alignment of those lenses which are involved in the magnification and projection of the image. Obviously this presupposes that one has suitable illumination reaching the specimen and the objective lens. This represents a rather separate problem. There are at least three elements to this system in all but a few simplified microscopes. The filament is the actual source of electrons, of course. The latter are given acceleration and direction by the applied voltage, and the shape and positioning of the hole in the anode cap. Then the condenser lens allows this beam to be modified convergently or divergently. When the beam leaves the condenser lens its axis must nearly coincide with the optical axis of the objective lens.

Two basic steps are necessary to align the illuminating system. The first may be spoken of as "source alignment," which is concerned with the alignment of filament tip, anode and optical axis of the objective lens. In essence the gun must be tipped so that its optical axis is parallel with that of the objective lens. The degree of tilt can be varied in most microscopes, either by actual mechanical tipping, or by controllable electrostatic deflection. Thus the axis of inclination can be placed on any meridian.

The second step in the alignment of the illuminating system is termed "condenser alignment." In essence the optical axis of the source, already parallel with that of the objective lens, is made to coincide with the latter. In the unaligned instrument the illumination sweeps, perhaps greatly, as the intensity (condenser) control is varied. In the aligned instrument the illumination expands symmetrically on either side of the cross-over image of the source. This is a condition of much practical convenience even though not of great importance in relation to resolution.

7.9. Source Alignment

Source alignment (at least in RCA microscopes) is best accomplished by taking advantage of an optical trick. One begins with a microscope lacking a specimen or objective aperture, but with a large, mechanically centered, condenser aperture in place. The objective lens current is greatly reduced (coarsest focus control), and the condenser set so that the source is viewed in focus (cross-over position). The beam current is then reduced substantially below the bias position, and "clouds" or "halos" are observed breaking away from the main beam as in Fig. 25. The position and character of these clouds provide the necessary information about the source alignment.

If two clouds, nearly identical in size and luminosity, break away symetrically from the ends of the cross-over ellipse as in Figs. 25(a) and (b), the source is properly aligned. However, if the clouds are quite unequal, or if there is but a single cloud at one end of the ellipse as in Fig. 25(c), there is misalignment in the direction of the long axis of the ellipse. If one or two clouds are seen at one side of the ellipse, as in Figs. 25(d) and (e), the misalignment coincides with the short axis of the ellipse. Of course intermediate relationships can exist.

In the unaligned position, the effective tilt of the system is in the wrong direction, and one must experiment with other tilts until alignment is achieved by successive approximation. In the older RCA EMU-1 and -2 microscopes, the gun tilt can be changed mechanically. In some foreign microscopes the filament is moved about within the anode cap which accomplishes the same thing. In RCA EMU-3 and EML microscopes, and in some

FIG. 25. Images of "clouds" or "halos" produced when the image of the source is focused at cross-over, and when the current is also below the bias position. The symmetrical cloud of (a) indicates good source alignment ((b) is natural size). Asymmetrical clouds (c)–(e) indicate misalignment which is corrected by successive approximation. The gun is tranversed in the direction of the main cloud, and then tilted in the opposite direction, as indicated by the arrows.

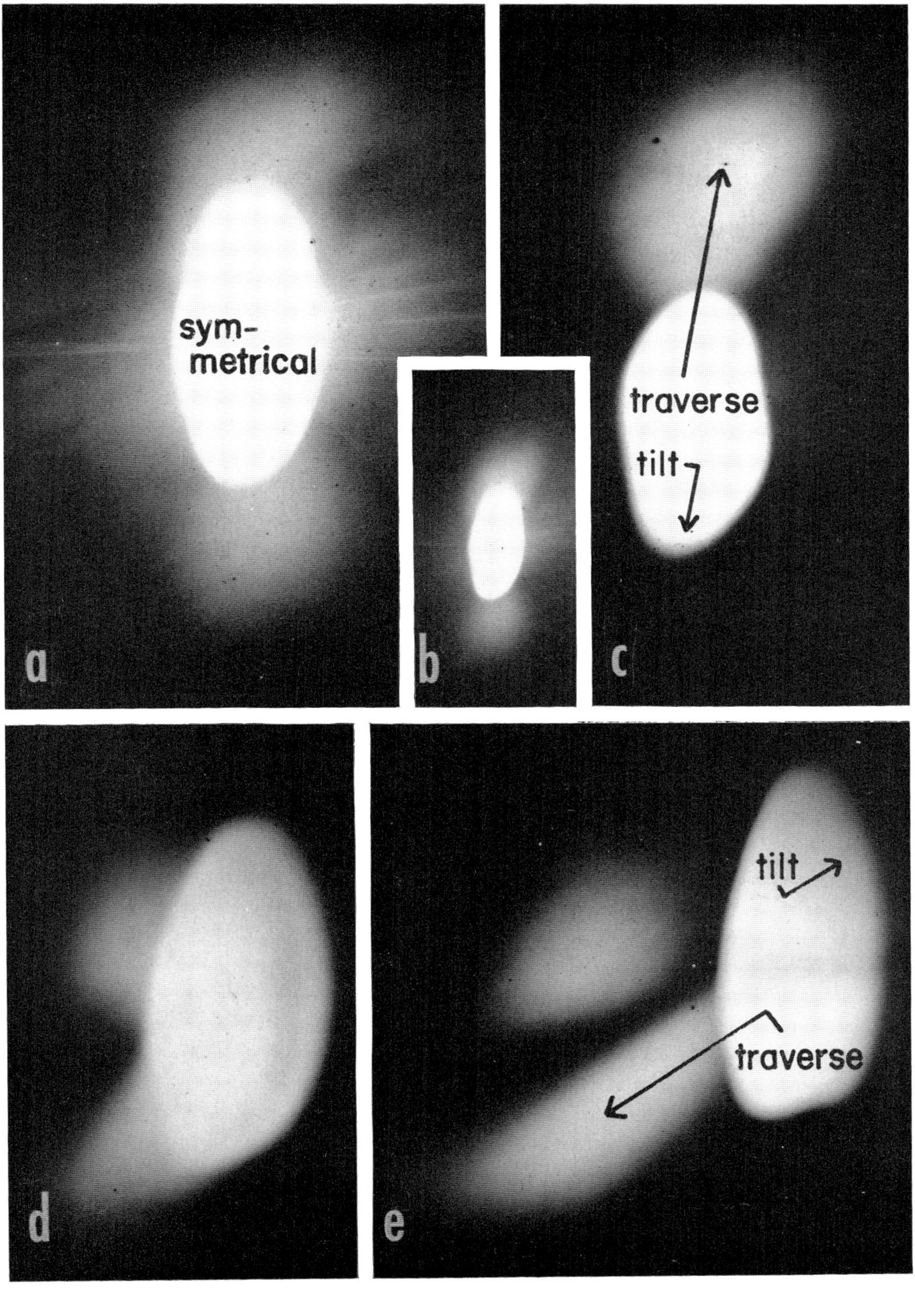
sym-
metrical
a
b
traverse
tilt
c
d
tilt
traverse
e

microscopes manufactured abroad, the equivalent of this is accomplished electrostatically.

To achieve source alignment, the gun is tilted so that the beam moves in a direction diametrically opposite to the asymmetric cloud. When the image sweeps off the viewing screen, it is returned by gun translation. In practice the source should be aligned at as low a magnification as is feasible so that as the image of the source is deflected by tilting, its path can be followed as long a way as possible. Also, one wants to see an image of the entire source on the viewing screen. As successive approximations approach alignment, the principal cloud dwindles in size as its counterpart increases, and becomes symmetrically disposed. If it seemed originally that one had a single lateral cloud, this will break up into two halos which will move to symmetrical positions at the ends of the long axis.

Source alignment is not very critical with a biased gun, and if it is obtained originally with a well-centered filament, subsequent replacement of the filament should not demand readjustment. Of course the tilt, once established, must not be altered. Centering the illumination between alignments must be done only by changing the gun translation.

7.10. Condenser Alignment

Condenser alignment is relatively simple. One is concerned with centering the gun in relation to the condenser axis. One must begin with a mechanically centered aperture, the larger the better to minimize centering error. As the condenser control is moved back and forth through cross-over, the beam will sweep across the viewing screen in an unaligned instrument. The sweep actually represents an arc of a circle, and one has to move the beam along a radius towards its center by condenser translation, while keeping the beam in view by gun translation. In practice, the arc is apt to have such a slight curvature that it is not apparent. It is usually simplest to choose a side arbitrarily and move the cross-over spot at right angles to the sweep. One soon learns whether the sweep is decreasing or increasing in amplitude. If the former, the correct side was chosen, if the latter, it is necessary to reverse the direction of the gun and condenser movements. Alignment is achieved when the illumination expands symmetrically on either side of cross-over.

Once the condenser lens has been aligned it should not need much attention. However, if small, externally centerable condenser apertures are being used, these will have to be put in after the basic alignment is complete, and will require alignment themselves. This is done once again by varying the condenser control through cross-over. The illumination will sweep until alignment is achieved when it will expand symmetrically. The aperture is moved in an arbitrary direction until the sweep pattern starts to change or illumination dims. In the latter case, one has been going in the wrong direction, and the movement must be reversed. When the sweep pattern begins to change, the aperture is then moved at right angles to the original direction, and the same criteria are used. Alignment is achieved by successive approximation, alternating the two axes of corrective movement.

7.11. Filament Replacement and Alignment

Filament replacement is the commonest maintenance task. Three different adjustments must be performed with some precision. (1) The tip of the filament must be mechanically centered with reasonable accuracy. If centering is done very carefully each time, there will be little need for optical alignment subsequently. Precision here pays dividends later. A centering jig which the manufacturer may supply or which can be made in a machine shop is quite worthwhile.

(2) The distance of the filament tip from the anode is critical in a biased gun. If it is too close, excessive beam currents will be drawn which are apt to overload seriously the circuits of the microscope, and destroy regulation. If the distance is too far, beam intensity will be weak. A depth gauge or jig which the manufacturer may supply or which can be made easily is almost essential.

(3) After the gun is reassembled, when the microscope is to be put back in use again, some alignment probably will be necessary, but no more than translational adjustments of the gun. If the previous filament was centered carefully, and the new filament is also well centered, no source or condenser alignment should be required.

Every time a gun is disassembled to change a filament, the operator should consider whether or not the anode cap should be cleaned. In the long run, it is apt to save time and effort to do this routinely with every filament replacement. As insulating contamination films build up around the aperture of this cap, bias relations are altered and beam current soon increases to dangerous levels. Furthermore, the insulating properties of these films establish aberrant, asymmetrical, electrostatic fields which distort the beam. The insulation is apt to break down from time to time allowing discharge. This causes flicker, which is often rhythmical, and can become severe. This leads to yet another consequence. If the objective aperture is at all asymmetrically contaminated, its electrostatic charge is sensitive to the amount of electron bombardment, and alterations due to beam flicker will cause image movements.

The anode cap can be cleaned with levigated alumina powder. The critical region is around the aperture, and particularly the side walls of the aperture. A finger dampened and coated with alumina powder is enough to clean the flat surfaces, but the walls of the aperture must be cleaned with a dampened cotton swab with a little abrasive. The alumina then must be removed by vigorous sudsing, followed by adequate rinses. The cap may be blotted with towels, but it should be wiped finally with a chamois skin or lens paper which is not apt to leave lint. However, before final assembly of the gun, the aperture area of the anode cap should be observed with a binocular dissecting microscope to be sure that no lint remains near the aperture.

7.12. Objective Lens Compensation

Every time a new objective lens aperture is installed the lens must be compensated. It must be emphasized that resolution cannot be better than the properties of the objective lens. Astigmatism resulting from aperture contamination is the big "bugaboo" which can appear rapidly and deteriorate images seriously. It is wise to check for astigmatism daily as one begins operation in the morning. Unless one has extreme confidence in his colleagues he should also check this after anyone else has used the microscope. When astigmatism appears it is essential that it be eliminated by compensation if it is not so severe as to require a new aperture. It is a "must" that you learn to use an externally controlled "stigmator" which should be considered mandatory equipment nowadays in any microscope used for section work. Either electrostatic compensation or its magnetic equivalent is acceptable. This is a feature of most new microscopes and is available as an accessory (manufactured by the Canal Industrial Corporation of Bethesda, Md.) for the RCA EMU-2 and EML instruments. The principle of the electrostatic method is described in Chapter 7.1.

Astigmatism can be seen most readily in diffraction images at edges. Thus, small particles suspended in space make satisfactory specimens. Most simply, one takes a bit of filter paper dipped in xylene, toluene, or benzene and ignites this in some safe place. A plume of dense smoke arises, and a new specimen screen is held in this smoke column until it is somewhat blackened, but not for such a length of time that its holes are completely obliterated by the accumulated carbon. Carbon particles obtained in this way are spherical except where they adhere one to another, and are small enough so that they can be observed conveniently at high magnification. Thus, diffraction rings appear fairly conspicuously when the image is slightly out of focus.

Small circular holes in fairly thick films are perhaps even better objects for visualizing diffraction rings (Figs. 26(d)–(m)). If one uses the method of section mounting that involves evaporating carbon directly upon naked sections, one will have plenty of small holes in almost any section screen. There will be no difficulty finding and selecting suitable small holes which will serve for compensation. One can even put aside particularly favorable screens from time to time, and keep them for compensa-

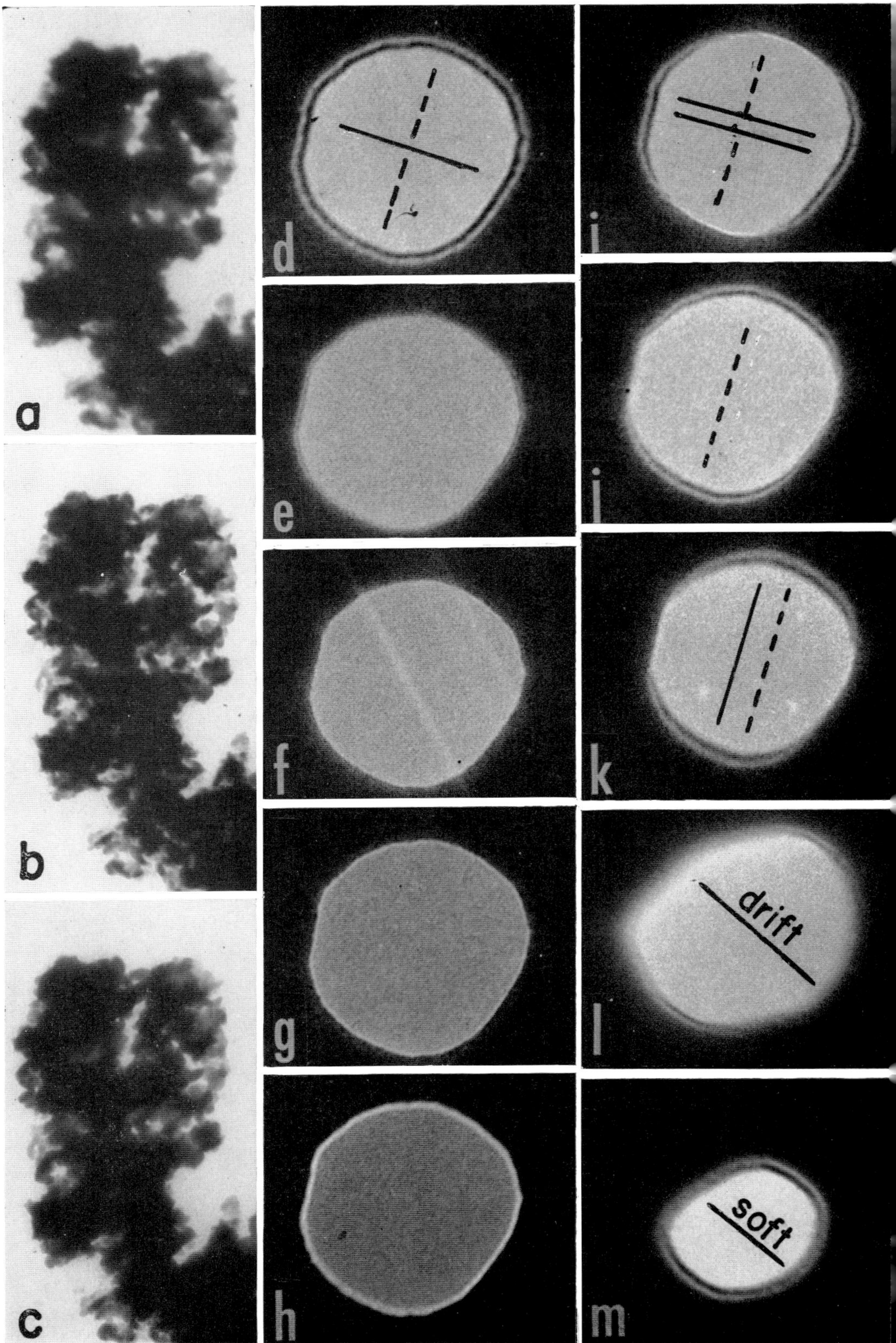
a
b
c
d
e
f
g
h
i
j
k
l
drift
m
soft

tion purposes. Thus, one is spared the necessity of making special preparations.

Chapter 6.4 describes how plastic films full of small holes can be prepared, which are eminently suitable for compensation purposes if they are given a relatively heavy coat of evaporated carbon. A group of grids prepared in this way can be made at one time and stored indefinitely. A single grid, in fact, can be used over and over again.

Compensation should be performed at the highest magnification step that one expects to use. With some instruments, compensation may be done at much higher magnifications, and it will be satisfactory in the lower ranges as well. This will not necessarily follow, however, and one should check this with his particular instrument and be sure that compensation is adequate

---

FIG. 26. (a)–(c) Overfocused, focused, and underfocused micrographs respectively of a complex dirt particle. The focused image shows much pattern complexity and a long range of tonal values. The out of focus images lack complexity and tonal variation. (d)–(h) A through-focus series of micrographs of a small hole in a film at high magnification; (d) shows the diffraction pattern of an overfocused image with a characteristic apparent light band outlined by black; (e) is very close to true focus with just a suggestion of overfocusing; (f) also is very near true focus but shows a slight rim of white suggestive of underfocusing. The bright band is more conspicuous in (g) which is definitely underfocused. An edge such as this has much greater apparent contrast than that of (e) and the novice is apt to think that this represents true focus when this is not so. In (h) the bright diffraction band of the decidedly underfocused edge is conspicuous and unmistakable.

The micrographs of (d)–(h) were made with a well compensated pole-piece using electrostatic compensation to counterbalance residual astigmatism. (j) shows the intrinsic astigmatism that existed in the lens system before compensation. (i) indicates how a relatively large, asymmetric electrostatic field can be superimposed, and its axis oriented at right angles to the axis of the original astigmatism. (The broken lines represent the axes of intrinsic astigmatism, the solid lines, axes of superimposed electrostatic force.) By reducing the intensity of the superimposed field, an exact balance of the opposing forces can be achieved to yield the compensated images of (d)–(h). In (k) the electrostatic force required to achieve compensation (as in (d)) was superimposed upon the axis of original astigmatism to produce an additive effect when compared with (j). (l) is the overfocused image of a hole in a film which was drifting when the photographic exposure was made. (m) is the overfocused image of a hole in a microscope showing "directional softness" (in this case due to an oscillating instability in the high voltage supply). Notice that the images of (l) and (m) are blurred at the poles of the axes of movement rather than being in-focus as are the poles of one axis of (i)–(k).

over the whole range. It is necessary to use a low power light microscope to see adequately the diffraction images on the fluorescent screen.

The over-focused diffraction image has a conspicuous light band outlined by a dark line which is fairly easy to see (Fig. 26(d)). With a well compensated pole-piece these bands circle the carbon particle or the rim of the small hole. It is symmetrical in all axes (Fig. 26(d)). The below-focus image looks rather different (Fig. 26(h)). The diffraction band is bright, and is not particularly visible since it does not contrast greatly with the illuminated background. However, it makes the apparent edge of the particle or the rim of the hole appear very sharply defined with great contrast. The in-focus edge (Figs. 26(e) and (f)), in which the true edge of the particle or hole is seen, will not appear with as much contrast as the latter. The edge will be sharp, it is true, but relatively indistinct.

When astigmatism exists the focal plane is different for two axes. Thus, if the poles of one axis are in focus, the poles of the other will show diffraction images corresponding to either the above or below-focus patterns (Figs. 26(i)–(k)).

One begins compensation by finding the axis of asymmetry using the over-focused image (Fig. 26(j)). One then imposes a rather larger and easily visible asymmetry on this field using the compensator controls (Fig. 26(i)). This field is rotated with the azimuth control so that this axis of asymmetry is at right angles to the intrinsic asymmetry (Fig. 26(i)). The imposed voltage is then reduced. The diffraction rings will be seen to approach the radial symmetry which is desired (Fig. 26(d)). If the axes of asymmetry have been established exactly at right angles to one another, no further compensation will be needed. More likely, however, some fine adjustments of azimuth and magnitude will be needed to obtain the truly symmetrical pattern of good compensation.

The difficulties that the beginner may experience in this procedure are mainly those of adequate visualization of the diffraction images. Illumination is apt to be faint at the high magnification where the compensation is done. Microscopes which have a double condenser system have an advantage in providing brighter illumination. Instruments which are equipped with a movable condenser aperture may be used with a large bore aperture in position to increase illumination while compensation is performed. In general, with experience and the knowledge of what is a suitable specimen, one can compensate with considerable accuracy simply by viewing the specimen screen with the low power microscope used in normal focusing. Certainly this is adequate for routine work with sections, although for very high resolution work, compensation may have to be checked photographically by taking a through-focus series of pictures.

7.13. Instrument Stability

Such a variety of things can destroy the quality of an electron micrograph that it is foolish to take a chance with the microscope itself. One should check its stability every day before use. This can best be done while one checks the objective pole-piece compensation (discussed in the previous section).

High magnification is used with an optical microscope viewer. It is simplest to superimpose specimen dirt on a piece of dust or a scratch mark on the viewing screen. Then any small movement can be detected. It may be that a drift will be observed that is due to specimen instability rather than to trouble with the microscope. This is particularly apt to be the case when pure plastic supporting films are used without carbon stabilization. In a micrograph images will be fuzzy in one axis (Fig. 26(l)). One can check this by examining rough contours of the specimen screen itself. These should not show drift for more than a few seconds after the stage has been moved, which should be sufficient time for thermal equilibration.

Most of the time, when there is image instability, the culprit is a contaminating film or a particle of dust in one of the critical places in the microscope column, and these should be considered and checked first. But it may be circuit instability! Most microscopes, unfortunately, do not provide ways of adequately testing circuit instability. The gauges that are provided are generally relatively sluggish voltmeters and ammeters. These do well enough to indicate the general level of circuit operation, but tell nothing of momentary circuit breaks, superimposed ac-ripple (Fig. 26(m)) or, in general, what radio technicians call "noise." In the absence of adequate test equipment for quickly spot checking circuit instability, the microscope operator cannot do much when image instability appears except first clean up the column. In at least nine out of ten instances this will eliminate the trouble. Since the objective aperture is so apt to be the cause of difficulty, it is usually well to examine image stability with this removed if one is suspicious of circuit instability.

It is an advantage to have a microscope more or less permanently equipped with built-in "power monitoring" equipment. Rapid voltage fluctuations, ac-ripple, and the like can be seen best on an oscilloscope screen. This writer knows of several instruments that have been custom equipped with an oscilloscope, suitably tapped in through switches to the important circuits of the microscopes. By flicking a switch, the behavior of any one of these circuits can be assessed at a glance. An "electron ray, indicator type" tube (6E5) can be substituted cheaply for an oscilloscope tube if used with a simple amplifier, and its rapid response to fluctuations by visible flickering will give most of the information that might be desired (Bishop, F. W. (1953) *Electronics,* **26**). The custom design, construction, and installation of such equipment, however, is apt to require the services of an electronics engineer, and most laboratories will not wish to go to the trouble and expense. The Canal Instrument Company of Bethesda, Md. is marketing a power monitoring system which is adaptable to several of the commercial microscopes. Although the system seems cumbersome and its usage overly technical for common checking, it will serve the person with some knowledge of electronics fairly well, and be quite useful in case of serious trouble.

Most laboratories, however, will not be prepared to explore electronic difficulties very deeply, and no doubt will call in the service representative when these arise. Before taking the microscope out of service, however, there are several things that the investigator should thing of and check, for he may be able to take care of the difficulty quickly himself. First, the sliding potentiometers attached to the control knobs get dirty and establish current leakage paths so that they cease to regulate circuits smoothly. A particularly common difficulty is to have an oscillatory make-and-break instability between one turn of the resistance wire and the next, so that a small circuit fluctuation is set up. This often is just enough to be quite damaging to image stability at high magnifications if it involves the focusing controls of the objective lens. When focusing is completed, one can tap the controls lightly with the fingertips to be sure that the image is stable. When the potentiometers become dirty in this manner they can be temporarily repaired by cleaning them with organic solvents and, perhaps, by using a fine abrasive cloth or paper. But the best solution is to buy new potentiometers which, in any case, will have to be done shortly.

Dirty switch and plug connections are in somewhat the same category as dirty potentiometers and may produce oscillating or irregular instabilities in the circuits in which they are involved. It is easy to turn panel switches on and off a few times to wipe their contacts, and it is not difficult to pull and reset column plugs two or three times to clean their prongs and sockets.

A common cause of electrical instability that is easily taken care of is battery "noise." In microscopes which use dry cell batteries for voltage reference, instabilities can be expected to occur as the batteries age, and often before there is much change in the rated voltage. The voltage instability is usually of sufficient magnitude so that tiny "kicks" can be seen on a standard voltmeter if the battery itself is tested.

As with any complicated electronic device, tube failure is an occasional cause of circuit failure. Tube failure, of course, can be a relative thing so that the tube functions to some extent, but below critical levels. It is not difficult to check tubes by substitution. Any laboratory that is at all distantly removed from a service source should stock a complete extra set of tubes. The investigator can substitute them in blocks. Thus, all those tubes connected with the objective lens circuit might be substituted at one time if this circuit is suspected. If the difficulty is corrected, the particular faulty tube can be determined by successive elimination.

7.14. Trouble Shooting

When an electron microscope behaves in an erratic manner a "differential diagnosis" is called for. We have already considered most of the common ailments in detail, but scattered through this and previous chapters. There follows a summary which, in essence, can be regarded as a check list.

(1) Image instability! This is most apt to be from dirt or contaminating films on the objective lens aperture or pole-piece. If the movements vary with beam intensity, these areas are particularly suspect. If it is the aperture that is involved, moving it slightly may give stability temporarily. Dirt or lint near the specimen, either on the supporting grid, in the bore of the specimen holder, or on the lip of the cap may be responsible. Then the instability is apt to vary as the stage is moved about. Contamination lower in the column, particularly in the projector pole-piece, is not uncommon. Electronic instability may be the cause. Dirty focusing potentiometers frequently are at fault. Poor contacts in switches and plugs may cause the difficulty. Tubes may be functioning inadequately. Dry cell reference batteries may be "noisy." Electronic difficulties producing image movements might be found in the high voltage, objective lens and projector lens circuits.

(2) Flickering illumination with a stable or nearly stable image! This indicates difficulty above the specimen level. The anode cap probably is dirty. This is apt to be associated with unduly high beam currents which may so overload the other circuits that electronic instability is present as well. A contaminated condenser aperture may be responsible, although these are so heated by electron bombardment that they tend to be self-cleaning. An unstable condenser lens circuit may produce

flickering. Some flickering can be tolerated for a while if it is ascertained that the image is stable in spite of this.

(3) Fixed sharp silhouette of contaminant! Objects protruding into the optical path below the level of the objective lens will appear in silhouette. They are most apt to be lodged in the projector pole-piece. They may rhythmically charge and discharge so that they oscillate.

(4) Extremely out of focus images of contaminants! Contaminants which protrude into the electron beam above or within the objective lens may produce vague images that are far out of focus. The objective and condenser apertures should be particularly suspect.

(5) Fixed image distortion! A contaminant below the level of image formation which produces an electrostatic field close to the optical pathway will produce a distortion of the image. The distortion will remain fixed as the specimen is moved about. The electrostatic field may be so localized, yet so intense, that images looking like silhouettes of globules may protrude into the field of view. These probably will seem to grow or diminish as the beam is altered. These are not true images of the offending particulate matter. Things introduced into the column such as a mechanical shutter or a photocell may give this sort of trouble.

(6) Particular parts of a grid image showing distortion! The distortion moves as the stage is moved. The contamination is on the specimen screen itself, the grid cap, or the specimen holder, and may be associated with image instability. The contamination may not be in the specimen plane, and so be completely out-of-focus. A piece of lint sticking through a hole in a supporting screen is a common cause. It may move about as the beam is varied and its electrostatic charge changes, and so the distortion also will move as the condenser control is varied.

(7) An unstable focus! When a high magnification image will not hold focus for a reasonable length of time, it is apt to mean the electronic instability of the high-voltage circuit. Reference batteries should be suspect, and also corona discharge or

other high tension leakage. High voltage is capable of jumping considerable gaps. So a circuit may actually be mechanically open and yet behave more or less erratically as if closed. The regulated power supply may not be functioning properly because of a faulty tube. Beyond these simple checks, a service call may be required unless the operator is quite skilled in electronic work. However, before deciding that the high tension is malfunctioning, it is important to rule out instability in the objective circuit, and even electrostatic effects from contamination in the objective polepiece which may at times simulate high voltage instability in their effects.

General References

Burton, E. F., and Kohl, W. H. (1946). "The Electron Microscope," 2nd ed. Reinhold, New York. (Useful mainly to the person without training in electronics or physics as a primer explaining basic theory.)

Cosslett, V. E. (1951). "Practical Electron Microscopy." Academic Press, New York. (Good history of the development of electron microscopy and descriptions of the early instruments. Excellent presentation of theory.)

Hall, C. E. (1953). "Introduction to Electron Microscopy." McGraw-Hill, New York. (Advanced presentation of theory and its practical application in detail.)

Wyckoff, R. W. G. (1958). "The World of the Electron Microscope." Yale Univ. Press, New Haven, Connecticut. (Simple presentation of electron microscopy, and an up-to-date survey of some of its accomplishments.)

8. Photography

8.1. Introductory Remarks

The electron microscope is fundamentally a camera, and the electron micrograph is its product. To achieve the latter we have the basic problems of any photography. Thus, this is one of the several skills in which an electron microscopist must acquire proficiency.

The first commercial microscopes in widespread use were, of course, the post-war RCA EMU-2 instruments. These were built to accommodate 2 × 10 in. plates which have proved to be a fortunate choice. This curiously sized plate was and is manufactured with a variety of fine grained emulsions, for it is of standard dimensions for such spectrographic work. The Eastman Kodak Co. "lantern slide" emulsion was recommended from the outset and hardly has been improved upon since.

The microscopist need not anticipate difficulty with commercially available photographic materials. His difficulties will lie in other directions. Most particularly he will find that focusing is a very serious problem, and also he will have to learn how to judge the image on the fluorescent screen so that he selects sections of suitable thickness and, in other ways, adequate for the task at hand.

It is obvious that only a small fraction of what the observer sees on the fluorescent screen of the microscope can be recorded on photographic film. Indeed, the time it takes to focus and otherwise adjust the microscope for picture taking, and to replace exposed plates, is a major production bottleneck. Inevitably, then, every electron micrograph of a tissue section represents a highly selected field. So much judgment is involved in the selection that it ordinarily cannot be delegated to technical help. The bacteriologist or virologist who has thousands of units all essentially alike might be able to do this. But the histologist who deals with ever changing patterns usually must expect to spend many hours looking at the fluorescent screen of the microscope, and to take dozens of micrographs before he can feel confident that he has understanding and adequate records of any particular specimen. This makes any comparative study very slow; for days and even weeks may be devoted to a single sample. It is apt to be difficult to make the uninitiated appreciate this. It is at the root of many criticisms that laboratories of electron microscopy are not sufficiently productive.

8.2. Section Thickness versus Magnification

To take low magnification pictures, one needs a different sort of section than when striving for very high magnifications. At low magnification, one wishes a large range of contrast in the picture. The "gray scale" should be so long that subtle variations in density can be seen. Relatively thick sections answer these requirements best, yet the sections must be thin enough so that one does not sacrifice resolution appropriate to the magnification (Fig. 27(a)).

At high magnification one usually is concerned with less complex pictures where a long gray scale is not important. One is content seeing objects in sharp contrast, black against white, almost as in an etching. Of course, one also needs great resolution. These requirements are satisfied by relatively thin sections.

In practice, then, no single section is apt to be optimal for both low and high magnification work. The problem is complicated further since some cells or organelles which are intrinsically very dense, or which have been "stained," will have

FIG. 27. Closely adjacent sections of the same arteriole as that shown in Fig. 28. (a) is of a section having very satisfactory thickness for a low magnification micrograph. Cytological detail is sharp and has good contrast. Such a negative can be enlarged until photographic grain interferes with image definition. (b) is of a section so thin that there was inadequate contrast for a low magnification picture. The negative was excessively "flat." In spite of adequate exposure, and even though this print was made on as hard a paper as possible, the reproduction is unsatisfactory. Such a very thin section might be useful for high magnification work, although one easily can have sections so thin that high magnification micrographs also lack adequate contrast.

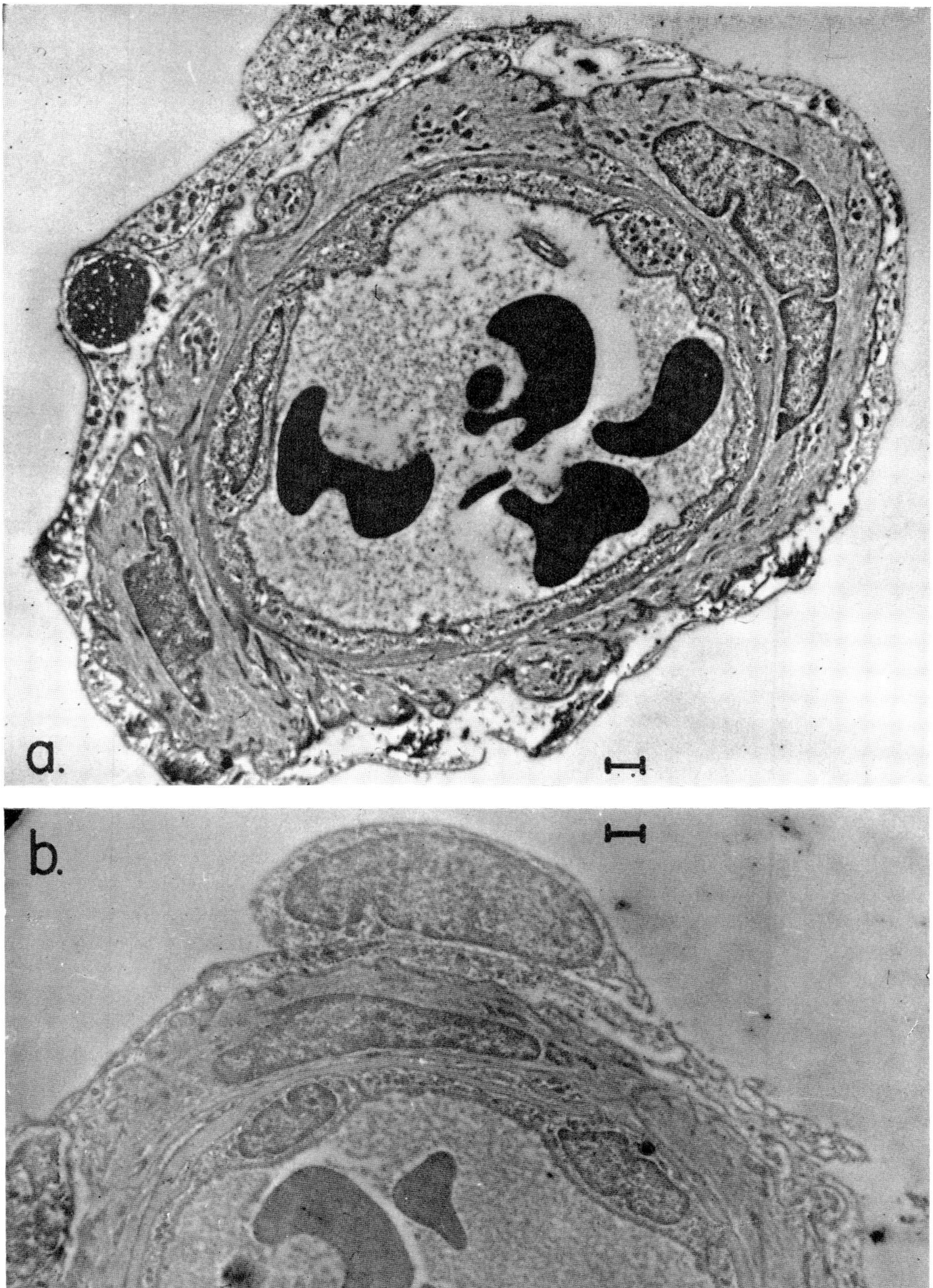
a.
b.

to be sectioned much more thinly than others of low intrinsic density to achieve the same final photographic contrast. This is well illustrated in considering a specimen such as a transverse section of a myelinated nerve fiber after osmium tetroxide fixation, in which the myelin and the axoplasm differ so greatly in density that both can hardly be shown successfully on a single micrograph.

The technique of ultrathin sectioning is such today that it is quite possible to cut sections too thin to make satisfactory micrographs. They may so lack contrast that they can be imaged only faintly on a photographic plate (Fig. 27(b)), even though photographic processes are exploited to produce the maximum contrast compatible with fine grain. If one needs to use such very thin sections for special purposes, "staining" methods in some instances may make it possible (see Chapter 5.22).

The microscopist will have to learn mostly by experience what images that he sees on the fluorescent screen will be useful on the photographic plate. The fluorescent screen does not have the contrast or resolution of the photographic plate, and the beginner will be lured into trying to take pictures of too thick sections (Fig. 28). He must learn to resist these, and photograph only those which have low contrast on the fluorescent screen.

8.3. Focus

The proper focusing of an electron image is surprisingly difficult for the beginner. The novice is apt to produce fuzzy pictures for some considerable time, and he is likely to blame all sorts of things except his ability to focus properly. The main difficulty is that the most vivid image with most apparent con-

FIG. 28. (a) A transverse section of a small arteriole. This section was too thick for good resolution (compare with Fig. 27). Such a specimen produces a high contrast image on the fluorescent screen of the microscope. The novice is tempted to record it photographically, but the resolution is so bad that one gains little or nothing by photographically enlarging it later. (b) Shows a portion of this section photographed originally at high magnification. Details are poorly resolved because of section thickness. The micrograph is quite worthless for serious work.

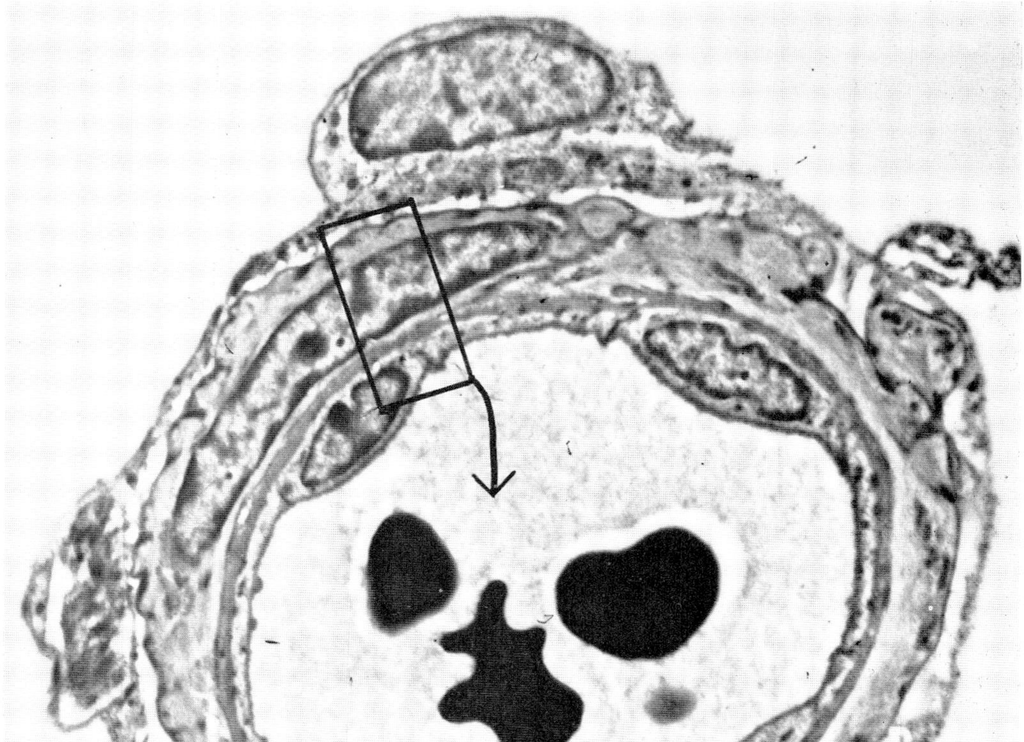
a.

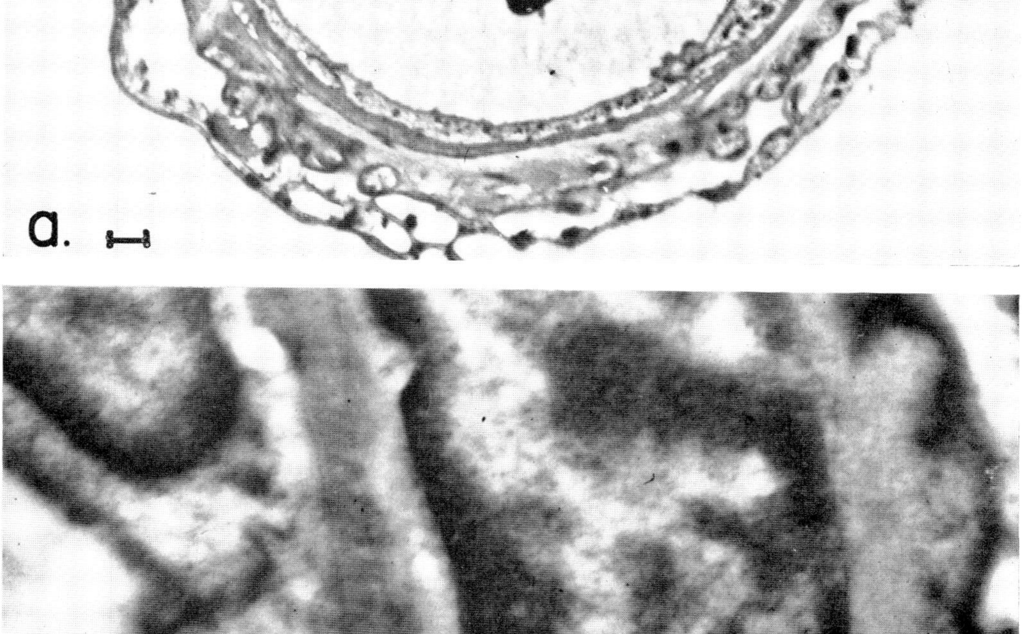
b.

trast ordinarily is a little out of focus (below focus). The heightened contrast of the out of focus image results from diffraction phenomena which are minimized in the true focal plane (Figs. 26(d)–(h)). The beginner must learn to focus intellectually, following rules rather than relying on his visual impression. The situation, of course, is slightly complicated because a fluorescent screen does not provide a truly sharp image. The photographic plate can be expected to show a wealth of detail not visible on the fluorescent screen. In essence we are able to record on a photographic plate detail that we cannot see to focus. It is this resolution that one strives for in making a picture.

It is foolish to attempt to take a picture if there is any doubt about the focus. One must resort to every possible aid. It is essential that one have magnification for viewing the fluorescent screen. This should correspond to the intrinsic resolving power of the screen, which will be about ×7–10. Binocular "dissecting" microscope bodies are ideal in this respect.

It is difficult, at best, to focus on the part of the image that one ultimately wants recorded. A part of a section that is thin enough to be useful finally is a very unsatisfactory object, for its contrast on the fluorescent screen is altogether too low. One needs a piece of dirt or a sharp edge for adequate contrast. Fortunately the electron microscope has relatively great depth of focus, so that if one can focus sharply on any part of a grid square, one then can move about within that square, and take a sharp picture of the desirable part of the specimen. It is the rare square indeed which does not have some little blemish that can be used for focusing.

The most accurate focusing of all can be done when diffraction bands can be observed at a sharp edge. For this purpose a hole or a tear in the section is ideal (Figs. 26(d)–(h)). The edge of a fold in the section or of an opaque particle may serve also. Some investigators deliberately spray a few latex particles on sections just so they can use their edges for focus. Out-of-focus diffraction bands are intrinsically narrow, however, so that this method is useful only at the middle and upper ranges of magnification. At low magnifications suitably chosen dirt particles serve best (Figs. 26(a)–(c)).

When out-of-focus diffraction bands can be seen, it is the overfocused band which is most conspicuous. Then the edge seemingly is outlined by a light line followed by a dark one; it is the light band that is particularly conspicuous (Fig. 26(d)). Below focus, the edge is outlined by a light band which gives heightened contrast to the edge, but since the background itself is light, this band is not conspicuous (Figs. 26(g) and (h)). The truly in-focus image shows no diffraction bands, and the edge therefore is without special contrast emphasis (Figs. 26(e) and (f)). If one examines an edge with high magnification and adjusts the fine focus control so that one goes back and forth through focus, it will be seen that at true focus the edge is least conspicuous. It is not difficult to approximate this by starting with the overfocused band plainly visible, and reducing the lens current until this just disappears. However, if the novice will take the trouble to make a number of test negatives he will soon realize that his eye is not really good enough to focus exactly. If true focus pictures must be obtained for very high resolution studies, a through-focus series of pictures is almost necessary. Fortunately, for most work, very slightly out-of-focus pictures are actually ordinarily preferable because of their heightened contrast; so that, in practice, one can rely on the fluorescent image for adequate focusing.

Naturally, diffraction bands will not be particularly evident if there are not suitable contrast conditions at the edge being observed. Thus, a hole in a very thin and transparent supporting film may prove inadequate. On the other hand, a similar edge in a section area where the density is greater can be expected to produce an easily visible diffraction band. If the edge of an object on top of a supporting film is being examined it probably will be necessary that the object have great density if a diffraction band is to be seen clearly. Thus, a fairly large, opaque piece of dirt may do, but a smaller, more transparent piece be inadequate. A fold in the section often will suffice.

Ordinarily the edge of a tear in a section that lies over an intact supporting film will not have sufficient contrast.

There will be times when the operator cannot use diffraction bands with high contrast edges for focusing. A square being studied at high magnification may not offer suitable opportunity for this type of focusing or, at low magnifications, the diffraction bands may be too narrow to be seen adequately. Then one must resort to another trick that ultimately also depends upon diffraction phenomena. One seeks out a piece of dirt, a small fold, or perhaps a thick ragged spot in the section. The object selected must show a range of contrasts. It must not be totally opaque or of a uniform grayness. It should have at least two distinct density values of its own which contrast with the relatively bright background (Figs. 26(a)–(c)). One then varies the focus with the fine control and observes the changing complexity of the image. The image is in focus when it shows the greatest variety of contrasts (Fig. 26(b)). In the out of focus images, contrast variation is lost so that the object comes to look uniformly dense (Figs. 26(a) and (c)).

Even a very small and inconspicuous contaminant may give all the information one needs for accurate focusing. In general, however, it will be found that the piece of dirt which is adequate at low magnification will not also serve at high magnification, and vice-versa. But it is the rare square, indeed, that does not have imperfections that can be used with reasonable success in this way. Occasionally, though, there may be no imperfections to aid focusing, and one must resort to the specimen in the section itself. The operator should then seek areas of greatest contrast. He will have to try to find a zone where there are reasonably contrasting lines or particles that can be seen well enough with the magnifying viewer to allow an estimate of focus. This may be difficult or even impossible at high magnification. If the operator cannot be reasonably sure of having achieved focus, there is no point in taking the picture. There are occasions when the most experienced operators simply have to abandon a grid square that they would like to photograph because they cannot find focus in it.

8.4. Special Considerations in Picture Taking

It is very important that the investigator learn to work fast in taking pictures, for contaminating films quickly build up on specimens. This tends to destroy resolution and quickly damages contrast. If one is beginning a new investigation and needs time to get oriented, it is best not even to attempt photography at first but use grids solely for visual observation. There comes a time when the investigator has gained confidence and has a good idea of what he wants to record on the film. Then, as he comes to each new grid square, he should decide without delay if anything is worthwhile recording. This decision must take into account section thickness and contrast which will determine whether high or low power pictures are feasible. No unnecessary time should be wasted in focusing and taking the picture. However, since it is useless to take an out-of-focus picture, all the time that is really needed to find a suitable focus object, and to be sure of accurate focus should be used.

Just after focusing has been completed, one should pause and be certain that there is no specimen drift which would ruin a picture. This is particularly important when unstabilized plastic supporting films are being used at relatively high magnification. They are then being subjected to an intense electron bombardment which produces strains as differential decomposition occurs. A moving specimen ordinarily should be abandoned, although very occasionally it will become stable if the beam current is reduced. It is particularly dangerous with unstabilized plastic supporting films to try to take pictures of squares which have the film split across the square, or which have pulled away from one supporting wire. Such films almost invariably show movement. The great advantage of stabilizing plastic supporting films with evaporated carbon is that the specimen *may* remain motionless even when quite asymmetrically supported.

When an intense (near cross-over) electron beam is being used there can be field effects which act like weak lenses. These produce a shift of the focal level. Thus, if the investigator focuses with a bright cross-over beam at high magnification, and then spreads out the beam to take a picture with much less intensity, he may have lost focus. This effect is perfectly visible with the naked eye if one looks for it. It is simplest to avoid the difficulty by focusing with the least possible illumination, and not changing this much, if at all, to take pictures. One can, of course, focus with a fairly bright illumination and then carefully observe whether or not changes occur when the illumination is reduced for picture taking. It is because of this potential danger of shifting the focal plane when illumination is altered that electron microscopists have come to take pictures at fairly high speeds in contrast to what was recommended originally. Most pictures are taken today at ½ or 1 sec (rarely over 2 sec). This corresponds to just enough illumination to allow accurate visual focusing. It becomes desirable and almost necessary to have an automatic shutter of some sort in the microscope. Moving the fluorescent screen as a shutter is too cumbersome, for unless this is done rapidly it is apt to expose one side of the plate proportionately more than the other. The automatic shutter should be activated by an automatic timer for working at these speeds.

Anything which improves the visual brilliance of the viewing screen, and so allows accurate focusing with reduced illumination, is advantageous. In some laboratories, the fluorescent screen is aluminized to produce a good reflecting surface, and then a very thin coat of fluorescent material deposited on top of this surface. There is, thus, back reflection within the fluorescent coat. The apparatus which is used for evaporating carbon will serve for this sort of metal evaporation.

The last thing one should do before taking a picture is to deliberately examine the field to be sure that it is evenly illuminated. It is all too easy to forget this, and to end up with a group of plates that have to be elaborately "dodged" in the printing process to compensate for inequalities in the original illumination. If one is working close to cross-over at high magnification it may, in fact, be quite difficult to find an adjustment of condenser strength and gun translation which gives truly even illumination.

8.5. Reasons for Unsatisfactory Pictures

Good electron micrographs are beautifully sharp in their details and have excellent contrast. If the beginner has trouble in achieving these ends in many of his micrographs he should consider the various possible sources of difficulty that begin when he first fixes the material, and end only when he has completed the finished prints. In assembling the following summary, some effort has been made to arrange these difficulties in order of frequency.

(1) Unduly thick sections. The beginner certainly will be tempted to take pictures of sections that are too thick to show good resolution (Fig. 28). This is easily corrected with experience when the operator learns to equate properly the contrast values of the fluorescent screen with those that will appear on the photographic plate.

(2) Unduly thick supporting films. Any supporting film reduces resolution and contrast so that the only good film is one which barely sustains the specimen. The operator should be willing to sacrifice many potential grid squares in order to achieve quality in the surviving ones. The beginner should be warned that a fairly thick section on a thick supporting film may look deceptively good on the fluorescent screen. It will have most of the contrast properties of a thin section on a thin film, but will, of course, lack the resolution of the latter.

(3) Specimen drift during picture taking. This was formerly a very important source of trouble. It need not be serious now if carbon films or carbon stabilized films are used in conjunction with fairly rapid shutter speeds, and if severely torn specimens are avoided.

(4) Changes in focus during preparations for photography. The beginner is apt to feel that he needs an unduly bright beam for accurate focusing. If the operator then reduces the intensity greatly, there may be a change in the focal plane of the ob-

jective lens so that the image he photographs will be, in fact, out of focus. There is less danger of this now than formerly because pictures are taken much more rapidly than originally, but every operator should carefully explore and master the characteristics of his instrument, and learn what the tolerances may be.

(5) Objective lens astigmatism. This limits the possibilities of obtaining high resolution pictures. It can be guarded against by making a daily practice of checking image quality.

(6) Microscope instability. This is most commonly caused by contamination of the objective aperture, but dirt or lint can easily get on specimen screens or the cap of the specimen holder and cause trouble. Furthermore, dirt can be carried into the column with the specimen holder and dropped down on the lip of the objective pole-piece. Of course there may also be circuit instability. Difficulties of this sort can be avoided easily if a practice is made of checking stability daily.

(7) Characteristics of specimen. Poorly fixed material frequently has little contrast. If it is simply a matter of starting with too large a block of tissue, there may be well-fixed material around a core of poorly preserved tissue. Thus, the same section may show a mixture of good and bad preservation, often with quite abrupt transitions. The beginner must learn to recognize as soon as possible what is bad preservation and not waste time trying to obtain good pictures of nearly worthless material. In some instances tissue components may have intrinsically such a low density that they will be difficult at best to photograph. Electron stains may become necessary, and are almost certain to be necessary when embedding is in something other than methacrylate.

(8) Sectioning artifact. Although most artifacts introduced by sectioning are easy to recognize, very fine order chatter may not be. It may tear apart the tissue on such a fine scale that the picture simulates poor preservation. One can imagine that this results from a scraping action of the knife rather than from proper cutting. The fault may be with the embedding material, knife angle, block size, or cutting speed.

8.6. Photographic Plates

Many investigators have been lured into a will-o'-the-wisp search for the ideal plate to use in electron microscopy. They, of course, have been seeking the ultimate in fine grain coupled with speed and a high contrast response. Their quest has sometimes rested on the false assumption that photographic emulsions respond to electron bombardment much as they do to light. While the photographic industry has produced technological marvels in producing a variety of films with a great range of response to light, no such range is to be found in their response to electrons. Differences in speed and contrast are relatively minor. But for once the electron microscopist has physical principles working to aid him rather than battle him, for the mere fact that electrons do not penetrate appreciable distances means that there is negligible lateral scattering within the emulsion. Therefore, image boundaries can be very sharp! This also means that the effective layer of emulsion is extremely thin, which intrinsically produces a great resolution as the photographic industry has lately discovered and is exploiting in films for miniature cameras. In effect, we have had this all along. Thus, the definition of an electron image turned out to be better than what might have been anticipated originally.

In general, an emulsion which is desirable for spectroscopy is also suitable for electron microscopy. The spectroscopist, however, often wishes panchromatic films which simply means unnecessarily complicated darkroom procedures for an electron microscopist. A slow orthochromatic film which can be developed with a bright safe light is all that we need. In the United States, the Eastman Kodak Company Lantern Slide Plate (contrast grade) has served very well. The company once published an amusing advertisement in various scientific journals apologizing for the unfortunate name that had been applied to this emulsion, years before it became useful to scientists for more than lantern slides. It is, in fact, a slow orthochromatic fine-grained emulsion with good, but not excessive, contrast characteristics. It is now spread on plates as a particularly thin layer so that there is a minimum of gelatin to be degassed before a good vacuum can be obtained. This is still the plate of choice, recommended by the manufacturer for most electron microscopy. It is a rare occasion indeed, when an electron microscopist can legitmately blame this plate for inferior quality in his work.

The Kodak Spectrographic Plate, Type V-O has a substantially finer-grained emulsion than the Lantern Slide Plate as a comparison of Figs. 29(a) and (b) will show. However, this is apparent only when the images are excessively enlarged photographically. The microscope and the nature of the specimen is more apt to limit resolution than the grain of the plate. The two plates have nearly the same contrast values, but the Spectrographic V-O Plate has only about one quarter the speed to electrons of the Lantern Slide Plate. This slow speed is a serious disadvantage.

The Ilford N.40 Process Plate and the Gevaert Diapositive Contrast Plate seem to have approximately the same emulsion as the Kodak Lantern Slide Plate, although this writer cannot speak from personal experience. For microscopes which use film, Kodak recommends their Fine Grain Positive Film, and the other companies have the equivalent.

Emulsions which are designed primarily for photoengraving can be used when extreme contrast is needed. Their use is to be discouraged for most histological work with methacrylate embedding for most gray tones would be lost and the micrographs be too harsh. They have their use in high magnification work where the contrast scale is narrow and the principal interest is in line. Therefore they are particularly useful for some

FIG. 29. A portion of a mitochondrion from an acinar cell of the parotid gland of a rat (× 165,00). (a) A micrograph made on Kodak Lantern Slide Plates, Contrast Grade and photographically enlarged excessively (× 11) to demonstrate grain size. (b) The same field photographed on Kodak Spectroscopic Plate, Type V-O. The original exposure on the latter plate had to be approximately four times that of the other. There were not, however, significant differences in the contrast values, and here both prints are on the same grade of enlarging paper. The spectrographic plate has a decidedly finer grain. However, photographic grain ordinarily is not what limits resolution, and in this instance the second picture does not reveal more detail than the first. Pictorial resolution was limited by the character of the specimen, and perhaps by the performance of the microscope.

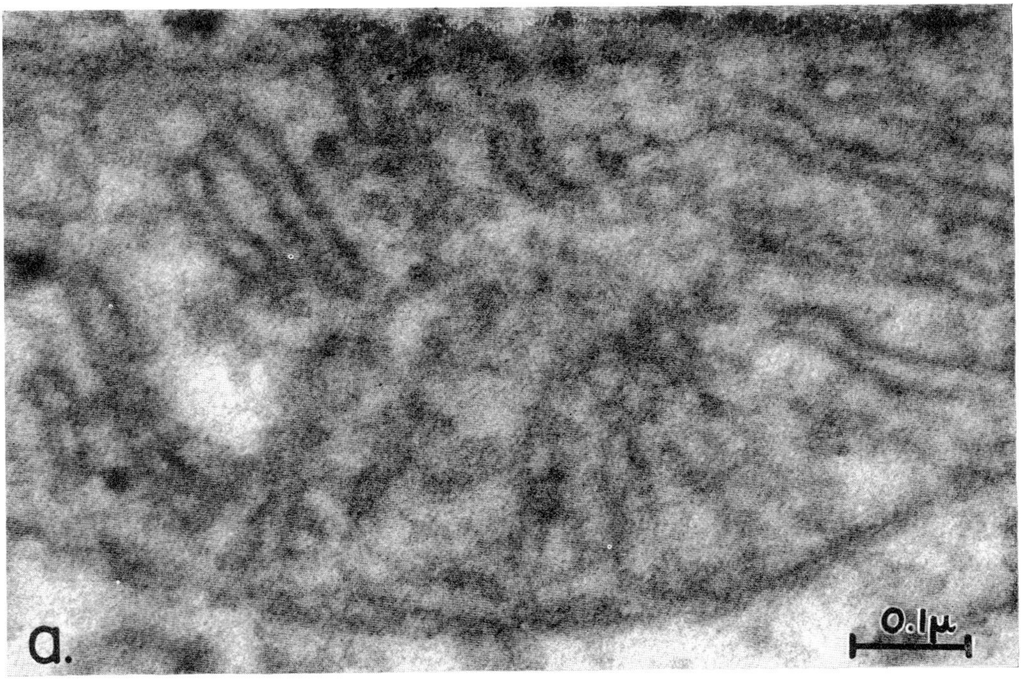
a.
0.1μ

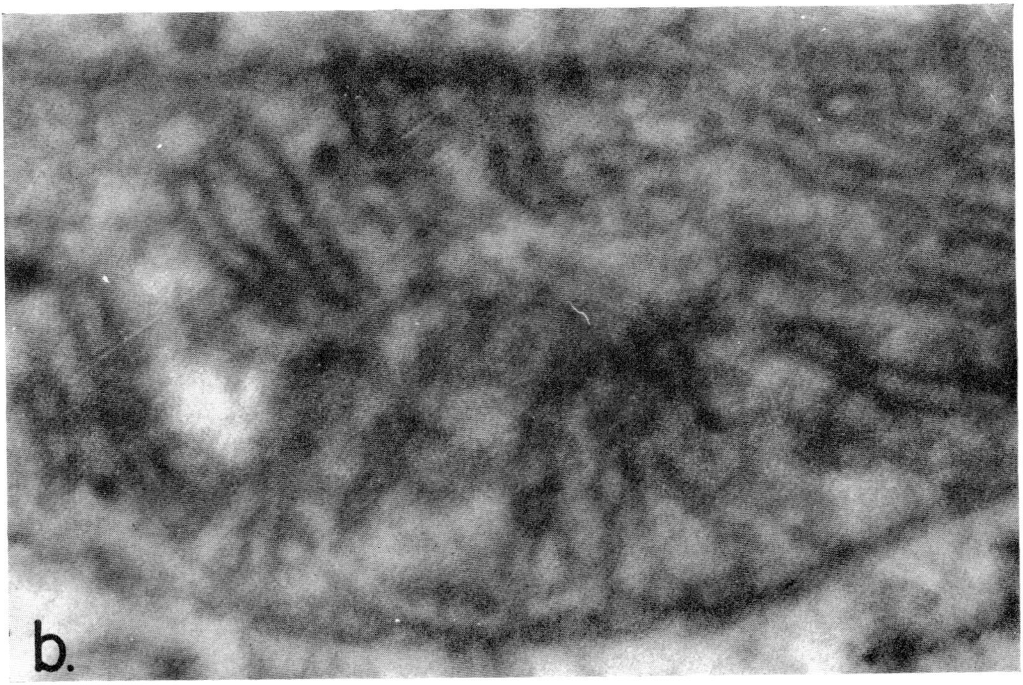
b.

types of macromolecular studies, and may prove valuable with Araldite and Vestopal W embeddings. In this category, the Kodak Company recommends Kodalith Ortho Plates, used with a special Kodalith, D-85 developer. The Ilford Company recommends their N.60 Photomechanical Plate, used with a special ID-13 developer. The Gevaert Company has similar emulsions on plates.

On the whole it is advisable for the individual investigator to settle down with one plate that he can obtain easily, and then learn to exploit its desirable properties to the fullest. His troubles will be compounded in a variety of ways if he chooses an excessively contrasty plate when his problems do not require it.

8.7. Plate Development

Almost as important as the type of plate used is the manner of its development if suitably fine grain and good contrast are to be obtained. In general, it is best to follow the manufacturer's recommendations, although with some qualification. Manufacturers usually suggest a choice of developers, one of which may particularly emphasize fine grain, another contrast, and a third convenience. Since contrast is often the greatest problem in electron microscopy, the developer which is selected should feature this. Unfortunately there is inevitably a reciprocal relationship between contrast and grain size with any particular emulsion. One cannot afford to choose the developer with greatest contrast without making too great a sacrifice in grain, but even less can one afford to sacrifice contrast for the finest grain developers. Truly fine grain developers lack contrast to such a degree that they are well nigh worthless for electron microscopy. Thus, one should select a developer tending to emphasize contrast, but with suitably balanced characteristics.

The Kodak Company recommends D-19 as the developer of choice for most electron microscopy. This was originally formulated for X-ray work, and later proved desirable for aerial photography. It gives a brilliant, vivid image with less "chemical fog" than most other developers. Range of tonal values is great.

Grain size is kept small. It is convenient in that it can be purchased already mixed in cans. It is a fast developer (3 min at 68° F or 20° C in a tray, 25% longer if used in a tank). It keeps well in a sealed tank and can be used with a replenisher.

Very high contrast developers such a D-11 are at times useful for special reasons, but, they have such a limited gray scale and their grain is so bad that they should not be used for most purposes. If necessary for particular purposes, a moderate increase in contrast can be obtained without sacrificing grain if plates are deliberately somewhat overexposed, and then given a short immersion in a reducing bath after development. Also "antifog" additives are available.

Investigators who work with 35 mm film in their microscopes have a special problem in that grain size does effectively limit the quality of their finished micrographs. There is no happy solution for them except to convert to larger plates where this is possible. This writer knows of several laboratories where Philips microscopes were being used which were pleased indeed with the immediate improvement in the quality of their work when the plate camera attachment became available and was first used.

Development is quite dependent upon temperature, and the manufacturer's instruction should be followed with reasonable care if a uniform product is to be produced. In particular, excessively high temperatures aggravate grain size seriously so that there may be days in summer when one should not attempt to develop plates, or at least temperature controlled water baths must be used with developing trays or tanks. Although modern developers (except the ultrafine grain ones) have considerable temperature latitude, this must not be exceeded. Plate development is best handled as a standardized operation varying only the time with the temperature.

8.8. Enlarging

Enlarging negatives to get the most out of them is a considerable skill which it is necessary to perfect. Many an excellent negative no doubt has been filed away forgotten because the print made from it was of poor quality. Some knowledge and several tricks are necessary to get the most from negatives.

It is quite important to standardize enlarging procedures as much as possible. A single developer such as Kodak Dektol (similar to D-72) should be chosen and used consistently, so also should a single series of relatively fast enlarging papers, limiting variation to the different contrast grades of one manufacturer.

Unfortunately, there is no standardization by different manufacturers in the way different contrast grades of enlarging paper are designated. Indeed, there is apt to be much variation in meaning in the products of a single company. In selecting a paper, then, one wants to be sure that a truly high-contrast grade is available. The Kodak Company's Kodabromide F series, with five contrast grades (numbered in order of increasing contrast from 1–5) serves well, and the number 5 paper is apt to be used most.

It is important to realize that contrast in a paper emulsion does not have the same meaning as in the case of a negative. All papers will produce nearly identical black tones and, of course, the "white" of these is the paper itself. Thus, every paper possesses the full potential gray-scale range, and proper printing spans the full range. Contrast in a paper, therefore, is not a matter of the range from white to black. It is rather a measure of the exposure required to produce the maximum black as compared with the exposure needed to produce a just visible tone. If the exposure needed to produce a maximum black is but four or five times that needed to produce the first visible tone, the paper has high contrast values, and is a "hard" paper suitable for printing "flat" negatives. The differential may be thirty times or more in a "soft" low contrast paper.

The implications of this are that the development of papers has relatively little effect on their contrast, since in all cases papers should be developed long enough to produce maximum blacks. Contrast control in the print must be governed by choice of paper contrast rather than by developmental control.

However, excessive overexposure and consequent underdevelopment will not produce a quality print. The full potential range of tonal values only appears upon full development, and the print which is "pulled" will lack these intermediate gray tones which add quality and, in electron microscopy, give useful information quite above and beyond their aesthetic values (Figs. 30(a) and (b)). Thus, the person without experience in photographic enlarging must learn that he will only get quality prints if he exposes correctly in the first place, and then develops fully.

The novice may wonder what is meant by a full development. The process of paper development has four rather well-marked stages which are easily recognized. There is first a considerable time interval after an exposed piece of paper has been put in the developer when nothing seems to be happening. Then, suddenly, an image begins to appear, and darkening of the major blacks occurs rapidly until the principal features of the picture are evident. There are not good tonal gradations, however, until towards the end of this phase as the development completes itself. During this time, development slows down so that the novice may think it is finished before it really is, and so he may remove the picture from the developer too soon. There is no danger in leaving a properly exposed print in the developer for much longer, however, for the third period is once again a time when nothing seems to happen. The fully developed picture can stay in the developer for 5 min or more before finally fogging starts, and all parts of the picture begin to darken. One should be sure that each picture reaches that plateau of inactivity which indicates full development if he wishes the maximum range of tonal values.

The beginner is apt to misjudge his printing in two respects. A wet print in a developing tray being observed with a safelight looks very much darker than it finally will appear to be. One has to learn to take this deception into account, and let prints become very dark looking indeed in the developer before fixing them. The beginner is also apt not to realize or to forget that pure white areas on a finished print perhaps should have had pale detail in them. Thus, if the print is underexposed, detail which was present in the negative may not be seen at all in the print. Professional photographers have a well founded rule that there should be no pure whites in a print. This is an excellent rule for electron microscopists. One should always expose so that at least a little gray tone appears in the lightest part of the print. This ordinarily can be seen appearing in contrast to the pure white edges that have been masked. It is only in this way that one can be sure to cover the full range of tonal values that are present in the negative. At the other end of the tonal scale, an enlarging paper of suitable contrast should be chosen so that only limited areas, lines etc., should be totally black in a properly exposed print.

The light source of the enlarger preferably should be spread over the negative by suitable condenser lenses rather than relying upon diffusion plates (or be a fluorescent source). In practice a "condenser type" enlarger produces an image with superior contrast, equivalent to about one contrast grade of paper. This is an important advantage.

8.9. Timing Print Exposures

As the previous section has indicated, it is essential that prints be fully developed if they are to have the quality that comes from a full range of tonal values. This in turn means that the enlarging paper must be properly exposed in the first place. There are electronic aids which can help determine the correct exposure (discussed below). These, however, are expensive and not generally found in darkrooms. Also, their use requires interpretation on the basis of knowledge, so that they hardly help the beginner. Most investigators will continue to use their eyes and judgment in estimating proper exposure time.

FIG. 30. (a) A properly exposed and fully developed print of glomerular capillary loops of a rat kidney, showing particularly an epithelial cell in the urinary space. (b) The same negative printed on an identical grade of paper, but "pulled." The negative was deliberately much overexposed and underdeveloped. Although the background contrast is much the same in the two pictures, the incorrectly printed one lacks tonal range.

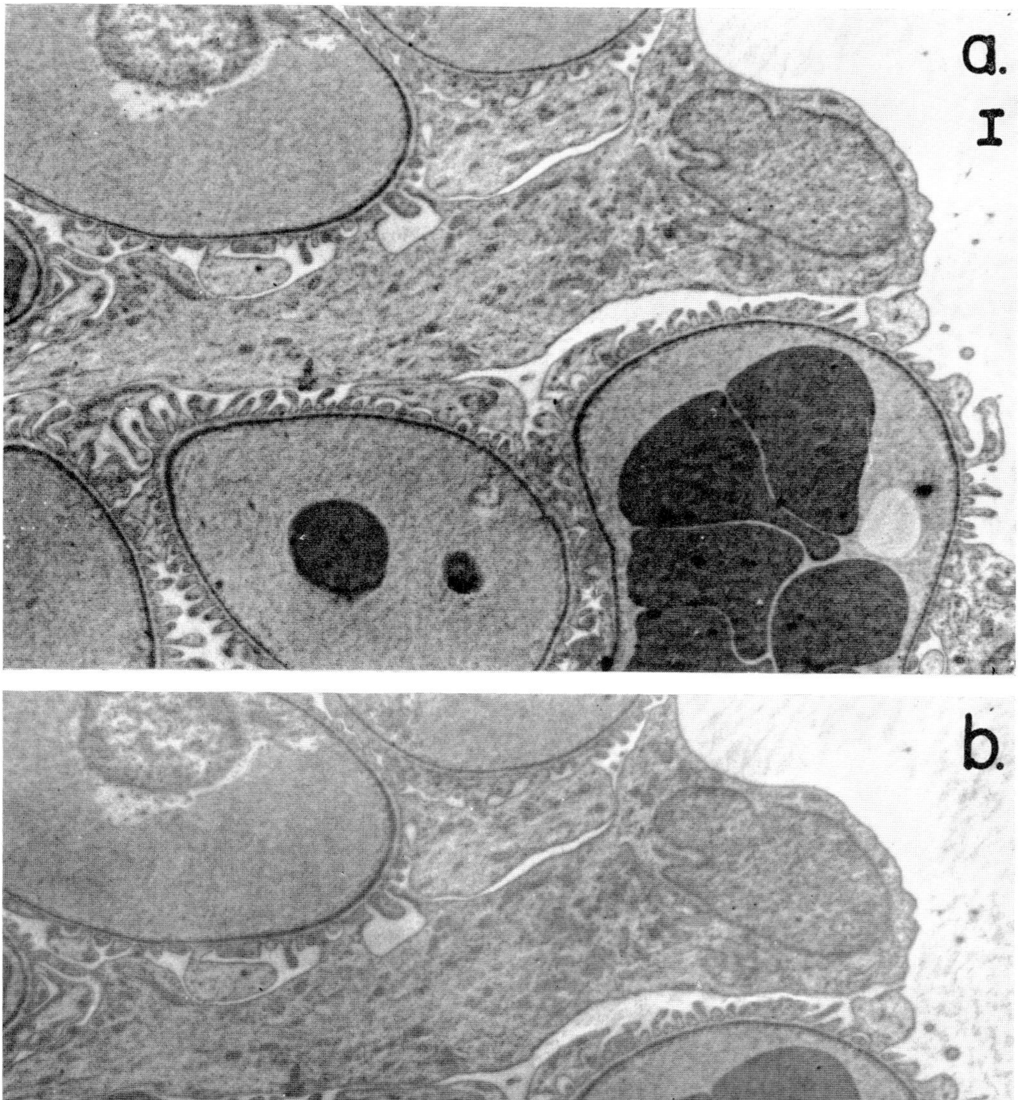
a.
b.

The beginner will certainly waste a good deal of photographic paper. This will be the cost of his education.

Every possible effort should be made to standardize the exposure. One has a choice of using a relatively dim image and long exposure, or a bright image and a short exposure. There are good reasons for deciding upon a fairly long exposure, perhaps 30 sec. The "stopped down" lens then will have excellent depth of focus. This also gives time for the manipulation of the image by "dodging" as discussed below. The eye can estimate the required image intensity fairly accurately in this range. With ordinary types of enlarging papers there is just enough illumination to visualize the image faintly. If the first exposure proves slightly incorrect, 7–10 sec can be added or subtracted with considerable accuracy, and produce just the right exposure. On the other hand, short exposures are harder to judge and afford little time for image manipulation.

The different contrast grades of enlarging papers require somewhat different exposures, the "soft" papers less than the "hard" ones. Therefore, one should make an effort to standardize illumination so that the most commonly used grade of paper is exposed for, let us say, 30 sec. If the next softer grade of paper is used, about 7 sec will have to be subtracted from this; if the next harder paper is used, about 10 sec added.

A densitometer with a suitable photocell pickup can be a most useful adjunct. The photocell should be in a small movable box, attached to the amplifier with a cable, so that it can be moved about on the enlarging easel and different parts of the image explored. There will be large fluctuations. One ordinarily determines correct exposure by what might be termed "average background density." If the background exposure is correct and the proper grade of paper is chosen, the rest of the image will take care of itself. The amplifier and meter, of course, have to be empirically calibrated beforehand so that one knows that a given meter reading corresponds, for example, to a 30-sec exposure with number 4 paper. The diaphragm of the enlarging lens is used to vary the illumination. Although such an instrument can insure that the basic exposure will be correct, experience and judgment is still necessary to select the proper contrast grade of paper and perhaps to "dodge" the image. SiSimpler "photometers" can be adapted for enlarging purposes, but their information is less accurate.

8.10. "Dodging"

The projection printing which one employs in enlarging offers considerable opportunity for controlling the image since not all parts need to be exposed uniformly. Areas of high density may be given an unusually long exposure so that they have gray tones on the finished print. Conversely, thin areas of the negative may be given brief exposures so that their details are not lost in solid black in the print. This is accomplished by "dodging" parts of the print with an out-of-focus mask.

The experienced operator usually can anticipate when an image will need dodging. He will plan his program in advance and decide, perhaps, that a central mass should receive 10 sec less exposure than the main picture. Then the central area can be blocked for 10 sec during the exposure by dodging it with a crudely shaped piece of cardboard fastened to a wire. This is held between the enlarger lens and the enlarging paper in an out-of-focus position so that the edge of the mask is completely blurred. Tonal values can be further blended by moving the mask a bit during the exposure. A corner or side area can be given additional exposure by masking the rest of the paper in an even simpler way. If a mask with a single straight edge is all that is needed, a piece of cardboard will suffice, or even the edge of one's hand. Angles can be made simply with two pieces of overlapping cardboard, or hands one on top of the other, or even the thumb extended away from the palm. If a central area of the image requires additional exposure, this can be achieved with a piece of cardboard having a hole in it, or even by letting a beam of light pass between overlapped fingers. It is desirable to have in the darkroom a few masks of different shapes that can be handily used as needed.

Although print control by dodging is an art important to

electron microscopists, unnecessary dodging wastes time, effort, and paper. If enough care is taken in making the original electron micrographs, dodging should not ordinarily be required. The commonest reason for having to dodge a print is carelessness in making the negative in the first place. In the excitement of focusing and adjusting the microscope to take a picture, it is all too easy to forget to be sure that there is even illumination across the area to be photographed. The image is particularly apt to be unevenly illuminated when one is working at high magnification with a close to cross-over beam. Rarely is the adage that "an ounce of prevention is worth a pound of cure" more applicable!

One other trick of print control has occasional usefulness. If a small portion of a print has been slightly underexposed, it may be "intensified" while in the developing bath by a little heat applied locally. This can be done by soaking the hands in the developer, and then squeezing the underexposed portion of the print between the palms so that close contact is made. Thirty to 60 seconds of body heat will darken the image somewhat, perhaps just enough to save the print.

8.11. Problem Negatives, Reduction and Intensification

Overexposed negatives can be saved if the overexposure is not too extreme. Any one of the standard "proportional" photographic reducers can and should be used. It is desirable to reduce them rather than to try very long printing exposures. If the negative density is such that a print exposure time of more than a minute is required, some fogging is almost certain to occur which will reduce contrast and print quality. Furthermore, in attempting to enlarge dense negatives, the lens diaphragm of the enlarger is wide open thus reducing its depth of focus, probably to the point where all parts of the image are not in exact focus. In cases of extreme overexposure it is best to abandon the negative for it will not reduce uniformly, and in all likelihood prints will be mottled.

In spite of the fact that some reducers are called proportional, the net effect is to heighten contrast in a negative that has been reduced. At times this may be an advantage, but it can also be a disadvantage if excessive. Thus, under most circumstances, reduction should be minimal.

Underexposed negatives are a more serious problem than overexposed ones. In general, if one cannot get a satisfactory print by straightforward enlargement on the hardest enlarging paper, one is best off discarding the negative. Attempts to salvage it are apt to be frustrating. The use of standard intensifiers is possible when only a small degree of intensification is necessary. They will not boost decidedly underexposed negatives to useful levels.

High contrast developers can be used with enlarging papers which help somewhat in improving contrast. Most packaged developers can be used without dilution, in a solution two or three times as strong as is normally recommended and, under these circumstances, will yield greater print contrast. "Antifog" agents can be used with developers which will heighten contrast a little, but produce unfortunate bluish tones. Such prints can be used for publication, however.

Contrast can be greatly increased by making a contact positive print of the underexposed negative on a lantern slide plate. Then yet another contact print, a second negative, is made from this. In each step contrast will be increased. One pays for this, however, in excessive grain, and the final negative is not apt to be useful if much enlargement is required.

8.12. Interpretation of Electron Microscope Images

It is easy not to realize or one forgets that electron images differ from photon images in very important respects. In a fundamental sense an electron micrograph is more closely allied to an X-ray picture than to a photograph. It is a shadowgram in which one records mass-density differences. Although one cannot ordinarily quantitate the tonal variations one observes, it is quite legitimate to use the information qualitatively.

This basic characteristic of the electron image has far-reaching consequences. For example, it is theoretically impossible to see sharply the edge of a wedge or a sphere suspended in space, for as one approaches the edge the density does not abruptly drop, but rather tapers gradually down to nothing. This has had more than theoretical interest for bacteriologists and virologists since their small organisms and particles presented a very serious problem in visualization until "shadowing" with heavy atoms was developed as a method of outlining them (cf. Chapter 9). It remains a problem when looking at particles and, more particularly, membranous structures that are embedded in the mounting medium of a section.

When a membrane is sectioned transversely, it appears as a line. When it is sectioned in the horizontal plane, it appears as a smudge. Oblique sections of membranes look like pale ribbons. The uninitiated may overlook the smudge, and interpret the ribbon as an out-of-focus image, or, perhaps, as an image spoiled by drift. Figure 31(a) shows at high magnification a greatly folded membrane. In many places the lines are sharp and obviously in focus. The rest of the micrograph must be also! Regions of apparent local fuzziness and lack of contrast thus require an explanation unrelated to the question of focus. Figure 31(b) of a small cluster of cilia, oriented in some-

---

FIG. 31. (a) Infoldings of the basal plasma membrane of cells of the distal tubule from a rat kidney at high magnification. In some places the membranes appear sharp and with good contrast, in other areas the same membranes become pale and lose definition. It is absurd to think that some portions of the picture are in focus while other portions are out of focus. The explanation lies in the geometry of the membranes. When they are viewed vertically in the section, their substance provides a maximum total density, and the boundaries are sharp. Remembering, though, that an electron micrograph is fundamentally a shadowgram, when the membranes are viewed obliquely or in the horizontal plane there is less scattering material in the path of the electrons and hence a reduced density. Also, edges tend to fade away in the image rather than being sharply delineated.

This section was "sandwiched" between two supporting films. This reduces the possibility of surface tension artifacts (see Chapter 6), but the second supporting film reduces the over-all contrast.

(b) A relatively thick section through a small tuft of cilia with slightly varying orientation (× 90,000). Those cilia which were vertical to the plane of section show considerable internal detail (*x*). Other cilia, less exactly oriented, appear blurred (*y*), although actually in sharp focus. The fuzziness would have been less evident had the section been thinner. (From the choroid plexus of a rat.)

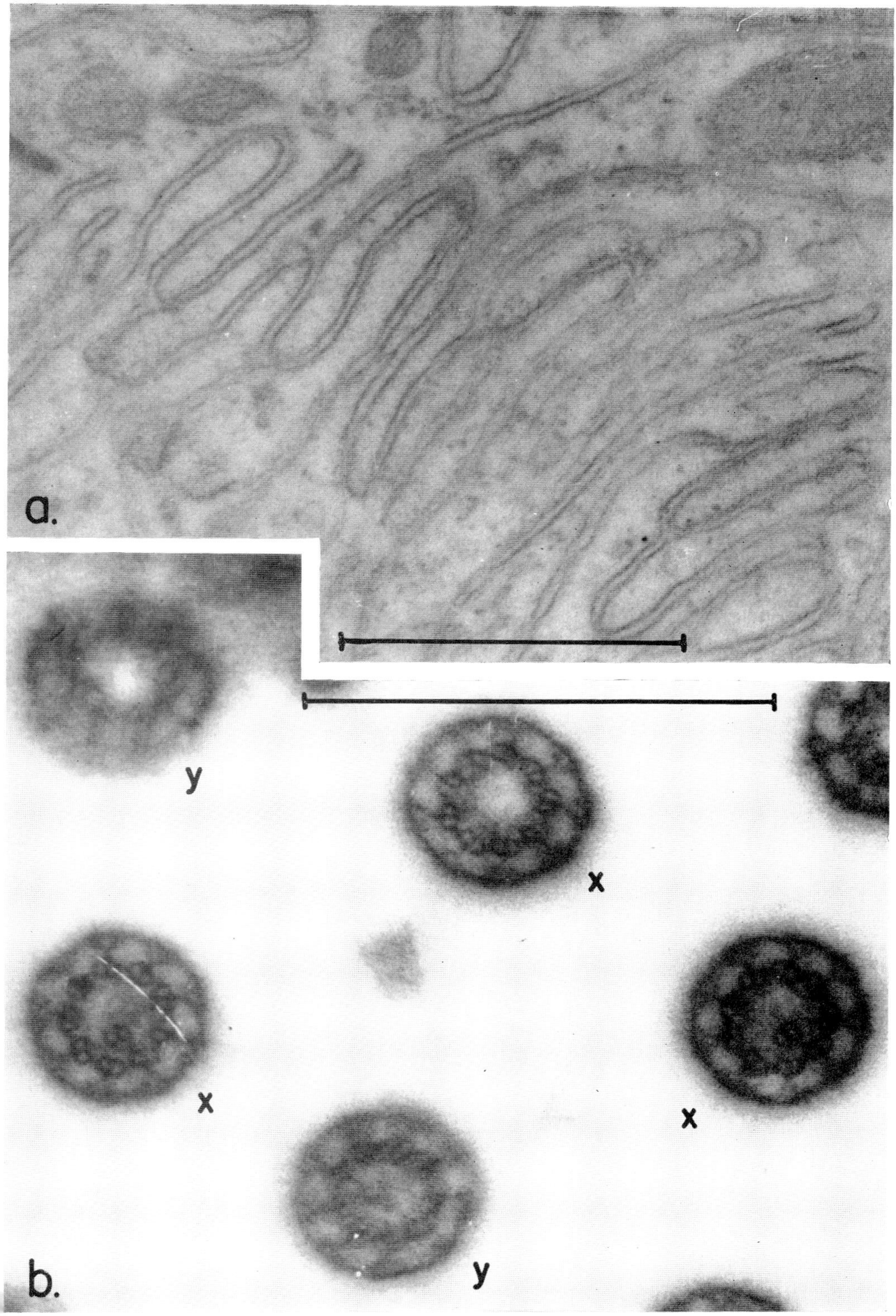
a.
b.
y
x
x
x
y

what different directions, also emphasizes the same point, for internal detail is quite apparent in the truly transverse sections (x), but the oblique sections (y) seem blurred. The mitochondria of Fig. 9(a) happen to have been oriented so that the cristae were in the plane of the section, and hence do not show. Thus orientation becomes a matter of considerable importance in electron microscopy, and it is often necessary to search diligently for a preparation which shows detail adequately.

General References

Lester, H. M., Photo-Lab Index, Morgan and Lester, New York. (A loose-leaf cumulative formulary of photographic materials and procedures. Quarterly supplements are issued to keep it up to date. New editions are issued almost yearly. This is the standard "bible" of the American professional photographer, but European materials are also included with considerable detail. It includes an extensive bibliography relative to photographic processes.)

9. Alternate Methods of Specimen Preparation

9.1. Introductory Remarks

There are, of course, some circumstances when it is desirable to examine as directly as possible dispersed or fragmented biological material. This is the common problem of the bacteriologist and the virologist who often have no need for ultrathin sections. Analytical work with cells and tissues sometimes yields particulate suspensions to be studied, and it is possible to isolate some fibrous components of tissue for direct examination. Furthermore, there are some hard tissue elements which almost defy ultrathin sectioning, but which can be studied by replicating their surfaces. The techniques which are useful for these operations include refined methods of fragmentation and suspension, the very delicate replication of surfaces, and shadowing with heavy metals to enhance surface contrast.

These methods had already been considerably developed before ultrathin sectioning proved successful. Replication techniques in general were not developed by biologists, but rather by metallurgists who were concerned with visualizing crystal structure in hard materials. The first successful replication usually is credited to Mahl in 1940. Substantial improvement, which gave much impetus to replication, followed the introduction of the Formvar technique by Schaefer and Harker (1942). With the introduction of shadowing methods by Williams and Wyckoff (1944) it was immediately obvious that contrast in replicas could be greatly enhanced by combining replication with shadowing. This has become almost universal practice. Shadowing, of course, also has been a boon to those who would study dispersed or fragmented material. This particularly includes the preparations of bacteriologists and virologists.

In the course of time, quite a variety of materials have been advocated as replicating substances. Some of these are plastics that require considerable heat and/or pressure to make the replica and therefore are not applicable to most biological work. Others are plastics or resins that can be deposited as films from solutions, and are more generally useful. Evaporated films, particularly of silica, are used. But when Bradley (1954) demonstrated that carbon could be evaporated easily *in vacuo*, it was realized at once that this was almost the ideal way to replicate most biological specimens. The chemical inertness of pure carbon allows most specimens or substrates to be dissolved after depositing the carbon, leaving the film isolated and cleaned. A carbon replica, thus, can be formed directly upon the material to be studied, or a thick plastic first replica may be used as an intermediate step if mechanical force must be used to strip off the replica. The face of this then is secondarily replicated with carbon.

It would not be worthwhile in this book, oriented as it is towards work with ultrathin sections, to attempt a full discussion of all of the variants of these methods. This chapter, therefore, will be limited to the most generally useful procedures for the biologist, and will only attempt to point out their main features and advantages. Often the interested investigator will have to consult the original literature for detail.

9.2. Suspended and Fragmented Material

The biochemists, of course, have various ways of suspending material which may then appropriately be studied with an electron microscope. Aliquots can be dried on supporting films and examined directly or after metal shadowing. Difficulties begin, however, when it is realized that there cannot be appreciable salt present in the suspension, or its crystals will coat the specimen particles and destroy their imaging. There is no solution to this problem except to design the experiment in such a way that salt is finally eliminated. It is also necessary to eliminate other contaminants, particularly those which might tend to form amorphous films which would envelope and bury the particulate matter in which one is interested.

Some fibrous tissue components can withstand considerable mechanical agitation and, therefore, can be cleaned as well as fragmented by the judicious use of blending equipment. This technique was developed by Schmitt and Hall and their co-workers for studying collagen (1942), and later striated muscle and other fibrous proteins. They started with commercially available electric blenders, but also used microblending equipment of their own design. Frictional heat develops quite rapidly within the suspension being blended so that running times must be kept short.

The simplest way of mounting suspended material is to place a drop on a grid previously prepared with a supporting film and allow this to evaporate. However, if the supporting film is hydrophobic, as it is apt to be, there may be some difficulties. Parlodion and Formvar films are mildly hydrophobic, and carbon films decidedly so. Carbon stabilized plastic films will prove most generally convenient and adequate for this type of work, but it sometimes is desirable to have the plastic on top of the carbon so that this is the exposed surface. To accomplish this, carbon must be evaporated on top of the plastic before stripping it from the glass slide. However, even traces of surface active material or wetting agents may be enough to allow drops of suspension to spread out over even hydrophobic supporting films so that difficulties will not be serious. In this connection, a thin layer of albumen deposited from 0.1% solution can be used.

If a drop of a suspension is simply allowed to evaporate on a supporting film, the suspension is apt to become concentrated as evaporation proceeds, and there may be much coalescence of material in the final stages of drying. Often this can be effectively avoided by employing a trick well known to the bacteriologists. A large drop with a high meniscus is placed on the film with no expectation of allowing all of it to evaporate. The drop is maintained for perhaps 30 sec or even a few minutes by the addition of more material if necessary. During this time some of the particulate matter settles and becomes attached to the supporting film. After a suitable time interval, as much fluid as possible is removed with a bit of filter paper and then drying of the residual traces of the fluid is permitted. One can expect that most of the suspended material will be fairly uniformly distributed over the grid.

Once organic material has dried on a supporting film, there is little likelihood of its washing off subsequently. Thus, one can employ "staining" procedures, histochemical reagents, or enzymatic digestions to such screens before examining them in the electron microscope.

The surface tension forces associated with drying tend to flatten particulate matter against the supporting film. (The degree of deformation ordinarily can be determined quite accurately after heavy metal shadowing by measuring shadow length and knowing the angle of shadowing in a manner described later.) Two methods have been advanced to circumvent these forces. The first of these is a simplified method of "freeze-drying" advocated by Williams (1953).

Williams used an atomizer to spray the suspension on a supporting film precooled with liquid nitrogen. This was done at atmospheric pressure, but within the apparatus that subsequently was to be used for the drying. Special precautions had to be taken to assure good thermal contact between the supporting film and the chamber walls.

After spraying, the chamber was evacuated and warmed to approximately —50°C. A cold finger near the specimen, however, was filled with liquid nitrogen. Relatively rapid distillation occurred from the frozen droplets of suspension to the cold finger so that drying was complete in about 15 min.

The drying chamber was opened only after it had been warmed above room temperature. The specimen then could be transferred to shadow-casting equipment. After shadowing, the supporting film was removed from the copper pedestal to which it was lightly attached by a "dry-stripping" technique and mounted on a grid. In this procedure the dried particulate matter falls down upon the supporting film where forces of adhesion ordinarily are sufficient to attach the particles. Very delicate structures may be preserved in three dimensions in this way.

Anderson (1951, 1953) developed the "critical point" method for eliminating the artifacts of surface tension. This takes advantage of the fact that the interface between a fluid in equilibrium with its own vapor disappears at temperatures above the so-called critical point. Moderate pressure liquefies carbon dioxide. Its critical point is at 31°C. Material can be suspended in and infiltrated by liquid carbon dioxide at room temperature if it has previously been infiltrated with amyl acetate following an alcoholic dehydration. Finally the bomb is warmed to about 45°C, and the carbon dioxide allowed to escape slowly. A dry three-dimensional specimen is left behind.

9.3. Shadow-Casting

To an electron microscopist shadow-casting means evaporating atoms of a dense material, usually a metal, from a point source at an oblique angle to the specimen. If the specimen is contoured, metal piles up on surfaces which face the source, but surfaces facing away from the source are shielded and receive no deposit. This is the "shadow" which in a positive print appears white. However, in a negative print the shadows appear black, while the metalized surfaces are light. Thus, negative prints do indeed look like pictures of contoured objects brilliantly illuminated from the side by light (Figs. 32 and 8), and ordinarily are used. Obviously, if one knows the angle of shadowing, θ, and the length of the shadow, l, the height of the object, h, which produced the shadow can be determined by simple trigonometry ($h = l \cdot \tan \theta$). In this way the third dimension of objects can be ascertained.

Shadowing is of great importance whenever small objects of relatively low density must be examined, for the contours of such objects may not be sharply defined otherwise. Furthermore, of course, fine order surface detail may be revealed which could not be visualized in any other way.

Metal evaporation in done *in vacuo*. The system must be

capable of a good vacuum if sharp shadows are to be obtained. The vacuum chamber should be readily accessible (usually a bell jar), and of considerable size (perhaps 10–12 in. in diameter and height) to allow flexible operation. Heavy gauge (approximately 15 mil) tungsten wire heated electrically vaporizes the metal used as the shadowing agent. For this one wants a low-voltage high-amperage system. A very heavy duty 6, 8, or 12-volt transformer which can handle currents approaching 100 amps (at least 50 amps) is required. Its output can be regulated by varying the input. Thus a variable transformer is used in the primary circuit.

The substance to be evaporated may be supplied in granular or even powder form, or be wire, ribbon, or foil. The tungsten heating wire can be adapted to accomodate any of these. For granules or powders small baskets are made by winding the tungsten wire spirally around the end of a pencil. If the metal comes in wire, ribbon, or foil form, it can be wrapped or clamped over the tungsten wire at a region where the tungsten wire is bent to a V-shaped point. The tungsten wire will heat most intensely where it has been kinked, and the molten evaporating metal will form a liquid drop at the base of the V.

A number of metals and a few other atoms can be used for shadowing. Chromium and gold have low boiling points and thus are easy to use. Chromium, however, is a fairly light element and therefore relatively thick metal layers must be built up to produce high contrast. This does not matter at moderate magnifications, and its convenience recommends it (Figs. 32 and 8), but it becomes a serious drawback if one is concerned with very fine order detail at high resolution. Also, partly because of the amount of chromium that must be employed, the images tend to have rather coarse grain. Gold is hardly ever used nowadays because of its tendency to aggregrate in large grains.

Nickel has a very small grain size but of course lacks great density. It tends to amalgamate with tungsten so that its evaporation is not reliable.

Paladium, and to a lesser extent platinum, were for a long

time the metals of choice for shadowing. Platinum is a little difficult to handle for it tends to form amalgams with tungsten if overheated. Although paladium will amalgamate with tungsten, its tendency in this direction is much less than platinum. Of course, these metals in being much denser than chromium, nickel or gold can be deposited in much thinner films. Their grain size remains small too.

In recent years uranium has become the metal of choice for much refined shadow-casting. Its density, of course, is very great so that very thin and therefore sharp films suffice to give contrast. Its grain size is small. When used for very high resolution work to bring out the finest order of surface detail, the film thickness is kept minimal, only a few atomic diameters thick.

For much work, great precision in the amount of metal deposited is not required, and visual estimation is adequate. A piece of white porcelain can be laid beside the specimen, at the same distance from the evaporating source, and the deposit observed as it builds up on this. Experience, then, will tell when the current should be shut off. When the substance to be evaporated is in the form of a wire, it is easy to standardize the deposit by cutting off and completely evaporating a particular length of wire, while keeping the specimen distance constant.

For very refined shadow-casting one can calculate the required weight of metal, m, for a film of given thickness, t (in Ångström units), using the following formula:

$$m = (4\pi l^2 td/\sin\theta)$$

in which d is the specimen distance and θ the angle of inclination. As an example, 15 mg of platinum would have to be evaporated to produce a 20 Å layer on a surface inclined 20°, 10 cm from the source.

The angle of shadowing is of considerable qualitative importance. Ordinarily one does not wish to have very long shadows unless quantitative measurements are involved when long shadows increase accuracy. Obviously one loses the detail of everything that happens to lie in shadowed areas. For particulate matter of roughly spherical proportions, a shadowing angle of 20° or 30° usually will be satisfactory. A moderately contoured surface may require greater obliquity, and 15° will be better. For very shallow contours, angles as small as 10° may prove useful. It is usually wise when possible to shadow simultaneously several specimen grids at somewhat different angles, and also at several distances, and select the most favorable one subsequently.

The distance between the evaporating source and the specimen is not critical. However, biological specimens that are too close to the filament heat can be burned. Also a long distance will make the source more effectively a point, and the shadows will be sharper as a result. Six inches is a convenient and safe distance to work at in most evaporating units.

9.4. Replication

Replication is used to study surface topography, which implies that one starts with a solid or semisolid specimen. Except when specimens are thin enough to be electron transparent there would rarely be a way of examining solid surfaces directly (although experimental reflection microscopes of limited resolving power have been built). One can, however, make replicas of surfaces which quite faithfully reproduce the smallest detail and are electron transparent. These exact representations of surfaces, then, can be studied almost as though one were looking at the originals.

Technical problems involved in making replicas, and particularly in isolating them from the original specimens, are quite varied. They often have been solved with much ingenuity. Thus this subject, even as it relates to biological problems, has many facets, and it seems necessary to subdivide it in the remaining sections of this chapter. The material has been selected to present a few methods in sufficient detail to be useful, and to suggest the range of variation and application. Types of replication and principles are discussed first, and finally replicating materials and the practical problems of handling them are considered.

9.5. Single-Stage Replication

If a replica is made in the form of a very thin film to begin with so that it will be transparent to electrons, it can be examined directly in an electron microscope. Such a replica is called a "single-stage" replica. If it is prepared by flowing a solution of plastic on the surface to be replicated, its exposed upper surface will be smooth, the other surface contoured. The replica itself will be of variable thickness, and will produce an electron image of varying contrast. Thus it can be used without further ado, and under ideal conditions is capable of providing tremendous resolution. As the replica film is thickened, however, contrast is quickly lost, and in most circumstances the replicated surface is observed more conveniently after metal-shadowing (as discussed earlier in this chapter). The contrast image produced by the shadowing is apt to overwhelm the minor density differences of the replica, and is the image studied.

If a replica is made by the *in vacuo* evaporation of carbon, silica, or some other light element, its exposed upper surface will be contoured much like the replicated one, and the replica will be of nearly uniform thickness. Such a replica must be metal-shadowed.

Inevitably, single stage replicas are quite delicate objects. They are apt to be difficult to isolate from the specimen unless the latter can be dissolved or digested away. If carbon is the replicating material, its chemical inertness may make this possible. If this can be done, they are capable of yielding beautiful results. Carbon films are quite brittle, however, and they must not be handled roughly. Usually they cannot be mechanically stripped from the specimen, at least not without a suitable backing film for additional support.

Plastic films differ from carbon films in being tough and somewhat elastic. Thus, sometimes thin enough films to serve as single stage replicas can be mechanically stripped or floated away from the specimen. (In some circumstances, of course, one might also isolate them by dissolving or digesting the specimen.) While very thin films often can be removed by stripping from metal surfaces, biological applications are pretty much limited to floating such films from the surface of glass, or freshly cleaned mica, upon which a particulate specimen had been deposited. Under these circumstances, the specimens can be expected to strip with the film, and have to be removed subsequently by solution or digestion if a clean replica is to be seen. The method has proved quite useful, however, in dealing with macromolecules, viruses, and the like.

The images produced by all single stage replicas are, of course, negatives of the original surface. However, this is not apt to cause serious difficulty in their interpretation.

9.6. Preshadowed Replicas

If a shadowed single stage replica is to be used, it is often best to metal-shadow the surface to be replicated first, and then build the replica film upon this. The metal will vigorously adhere to the replica, and usually can be removed with it. Such replicas are properly called "preshadowed replicas," or "pseudo-replicas." The resolution of such replicas is limited only by the "grain" of the deposited metal.

In this procedure one ordinarily is concerned only with the image produced by the metal-shadowing, and pays no attention to the relatively faint image resulting from thickness variation of the replica. For this reason the procedure more properly should be called a "shadow-transfer" technique.

The method has found much application in virology and macromolecular research. It has an advantage over a single stage replica which is shadowed after replication in that the shadow images are positives. Also, the angle of shadowing can be known exactly and employed quantitatively.

One interesting but highly specialized application of the method is the technique developed by Steere (1957) in which frozen material is first planed *in vacuo*, then etched slightly by vacuum sublimation, shadowed with chromium, and finally covered with a replicating film of evaporated carbon. Adhering biological material is later removed from this preshadowed replica by boiling it in concentrated sodium hydroxide. If the original specimen is a frozen block of tissue with a shaved surface, the final image will be that of a "pseudosection."

9.7. Two-Stage Replication

Single-stage replicas inevitably must be delicate objects, and are only practical when a thin film can be stripped easily from the surface to be replicated, or when the specimen can be dissolved or digested leaving the replica behind. One can, fortunately, employ "two-stage" replication to avoid these limitations. Then a suitably thick primary replica is made first which can be mechanically stripped from the specimen surface. This surface is in turn replicated to produce a second and final, thin, positive replica. The primary replica is removed by differential solution.

It is obvious that the final replica may be either preshadowed to give a negative image of the specimen surface, or shadowed as a final step to give a positive image. Since preshadowing avoids distortion, it is more commonly employed.

Nowadays evaporated carbon is almost always the material of choice for the final replica. Its insolubility in almost all reagents allows it to be isolated and cleaned. Primary replicas are most commonly made of Formvar.

Two-stage replication is advantageous when it is desirable to remove the replica by mechanical stripping and when the specimen is large enough so that these operations can be performed under a binocular dissecting microscope, or even without optical aids. Hard tissues such as bone, teeth, and even hair can be handled easily (Fig. 32). Soft tissue surfaces which can be dried, such as skin, do not present difficulty. There have been many botanical applications of the method.

9.8. Preparation of Surfaces for Replication

Surfaces to be replicated are apt to require some preliminary preparation. If one is interested in mineral patterns as, for instance, in bone or tooth, mild etching with weak acids may re-

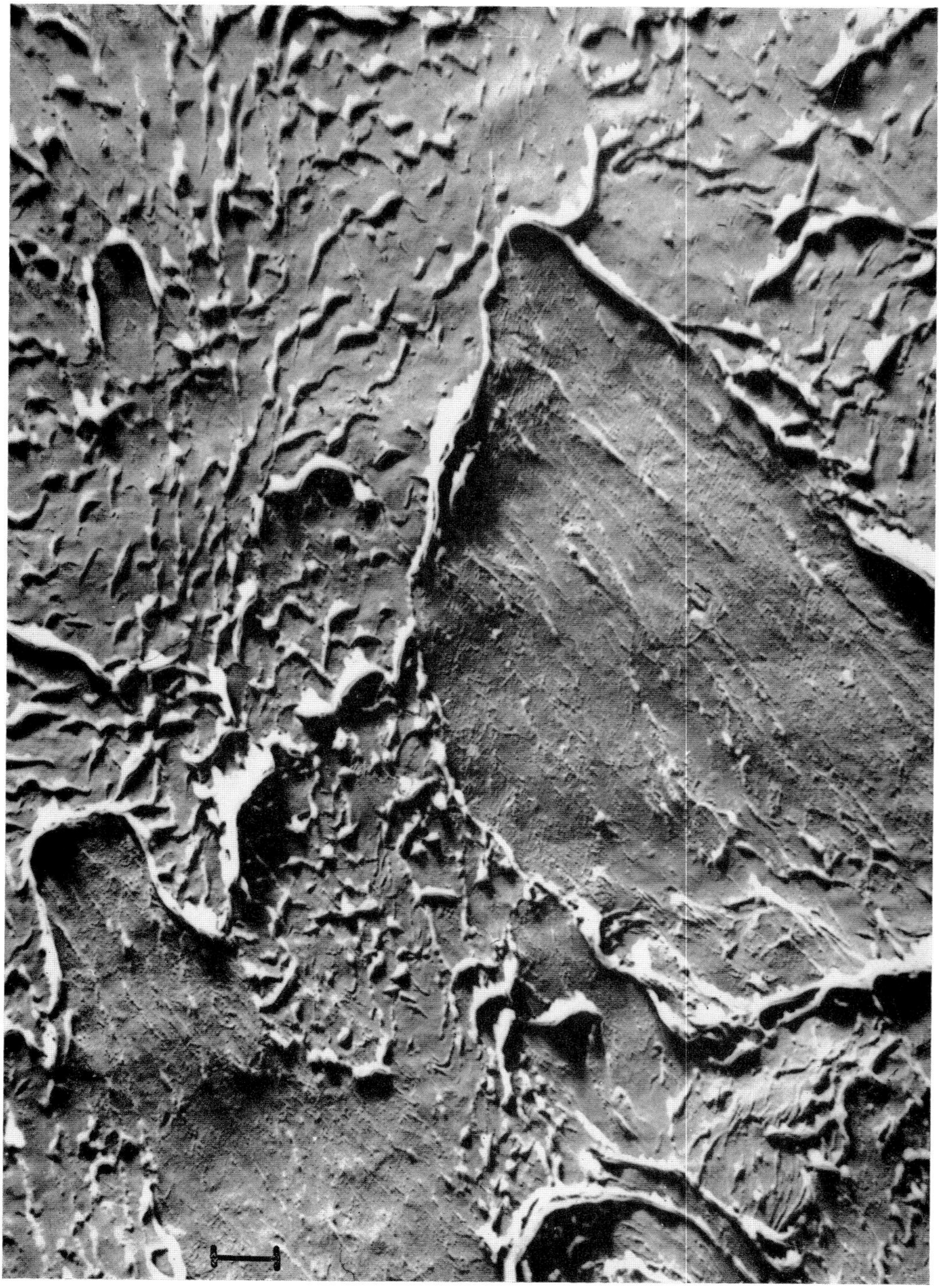

veal detail not otherwise visible, or freshly cleaved fractures uncontaminated by handling may be the surfaces of choice. Metallurgists frequently begin with a polished surface which is then etched. Surfaces like skin or hair should be degreased using suitable organic solvents (Fig. 32). Paint and wood surfaces have been studied after differentially extracting various components.

If a primary replica is to be made by a plastic such as Formvar dissolved in an organic solvent, it is important that the solution wet the surface completely and not trap bubbles. Ordinarily the specimen should be wet with pure solvent first before flowing on the solution. If a very volatile solvent such as chloroform is being used, care must be taken to avoid the condensation of water droplets.

If a primary replica is to be made by the vacuum evaporation of carbon, or some other substance, the surface of the specimen must be extremely clean. The use of organic solvents as cleaning agents may be necessary, but is particularly hazardous since trace quantities of nonvolatile impurities can so easily leave contaminating films that will leave their mark on the replica, and quite likely bury the detail one seeks.

9.9. Materials for Replication, Deposited from Solution or Melt

Surfaces can be replicated excellently by a variety of substances deposited from solutions including particularly plastics and resins. In this category, solutions of Formvar in chloroform or ethylene dichloride have been most commonly used. Parlodion in amyl acetate behaves similarly. Thin films, suitable as

---

FIG. 32. The surface of the lunula of a fingernail. A fairly thick Parlodion replica of a freshly exposed surface of the fingernail first was made, and then obliquely shadowed *in vacuo* with chromium. This surface was secondarily replicated by evaporating carbon upon it from several directions and, particularly, vertically. Subsequently the Parlodion was dissolved away leaving the thin carbon replica for microscopy. This process is properly called two-stage replication. The first replica is a negative, the second, a positive. The second replica can also be spoken of as being preshadowed. This is a negative print, as is commonly done with shadowed material, so that the "shadows" appear dark rather than light.

single-stage replicas, can be formed by flowing dilute (approx. 1%) solutions of these substances over the surface to be replicated, draining and drying. Thick films, suitable as the first replica in two-stage replication, are formed if viscous, concentrated solutions are used. It is also possible to thicken the replica by building up successive layers of plastic.

A promising, recently described method for a thick first stage replica of a flat surface starts with a thin sheet of polymerized methacrylate which can be purchased commercially. This is soaked in methacrylate monomer until its surface is softened. It is then laid over the surface to be replicated, covered with a piece of glass and squeezed moderately. Apparently there remains sufficient catalyst in the original sheet of plastic to polymerize the new surface layer in a period of a few hours, or overnight, without resorting to heat.

Other replicating substances which can be deposited from organic solution include resins such as "Zapon," and the nitrocelluloses and related compounds. Water soluble substances include polyvinyl alcohol, methyl cellulose, and gelatin. If the specimen can withstand moderate temperatures (from 45° to 80°C), replicas can be made using melts of many resins and even waxes. The metallurgists, who ordinarily are not concerned with temperature limitations, like to use thermo-setting plastics such as polystyrene with applied pressure as well as heat.

Under some circumstances plastic replicas can be isolated and cleaned by dissolving or digesting the specimen. If this is to be done nowadays, however, evaporated carbon (discussed in the next section) more likely is the material of choice for the replica. Plastic replicas are ordinarily used when they are to be floated off the specimen, or are to be mechanically stripped with force.

The character of the specimen will determine whether the relatively delicate flotation method is possible. If so, it generally will be necessary to tease the edges of the replica to get it started. Breathing on the replica to coat its surface with condensed moisture sometimes helps. There will be many specimens, however, which prove completely refractory to all such tricks.

Mechanical stripping will require little or much force depending upon the nature of the specimen. This, in turn, will determine in part how thin a replicating film can be obtained. Stripping may be aided by combining it with flotation, running a drop of water between film and specimen as they are pulled apart. Sometimes a preliminary period of mild oven drying helps; so may chilling in a refrigerator! The bonding between specimen and replica may also be altered by time, so what is impossible immediately after casting may prove possible the next day. No standard rules can be given, and each new problem has to be approached with ingenuity, but with the knowledge that "tricks" may help.

Some replica films will be big enough and strong enough to grab with forceps and pull. But more often a more delicate method of stripping is required. Adhesive cellulose tape commonly is used. This is rubbed over the replica to assure good contact. After stripping, the replica is separated from the tape with a suitable organic solvent that does not affect the replica (i.e. acetone if the replica is of Formvar).

If single-stage replication is being attempted with a plastic replica film that must be stripped, it is usually desirable to transfer it directly to specimen grids. This often can be done by laying grids upon the replica, and protecting their upper exposed surfaces from the adhesive tape with interposed bits of lint-free paper (cigarette paper serves well). When the grids subsequently are pulled away from the adhesive their centers will be clean. This method of transfer of course is hazardous, for too delicate a replica film will tear in the vicinity of the grid and not strip there. The bonding strength of replica and specimen often will be so great that there is no possibility of transferring a thin enough replica to serve as a microscope specimen.

Films suitable for single-stage replication sometimes can be reinforced for stripping by coating them with other substances of different solubility. Gelatin is easy to use, and thick films can be built up from heat concentrated solutions upon Formvar or Parlodion replicas. The dry-stripped replica finally is cleaned for mounting by soaking it in hot water. Polyvinyl alcohol in water and carboxymethyl cellulose in water can be used similarly. The very different solubilities of Formvar and Parlodion allow one to be used to reinforce the other. Flexible resins such as Zapon can be used. Various waxes have been employed.

9.10. Materials for Replication, Deposited by Vacuum Evaporation

It was early realized that replicating films might be built up by the vacuum evaporation of light elements. Silica (ordinarily produced by the evaporation of silicon monoxide which vaporizes at a lower temperature than the dioxide) yielded good results because of its chemical inertness which allowed relatively easy cleaning, and also because of its stability in the electron beam. However, with the advent of simple carbon evaporation (Bradley, 1954) the technique has been largely supplanted.

Carbon, of course, is a light element and therefore relatively electron transparent. It is chemically quite inert which allows films to be cleaned easily with almost any organic or inorganic solvent. Thin films are tough and stable under electron bombardment. There is only one serious difficulty with carbon; its films are brittle!

The brittleness of carbon practically precludes stripping replica films from a specimen. Therefore carbon can be used only when the underlying material can be dissolved or digested away. If a single-stage replica of particulate material is being sought, it is often possible to deposit the sample directly upon specimen grids previously prepared with Formvar or Parlodion supporting films. Usually these would be metal shadowed first, and then a carbon replica deposited on top. To insure a complete carbon film it is best to evaporate it obliquely from three divergent angles, or rotate the specimen *in vacuo* while deposition is proceeding. Finally the plastic is dissolved away, leaving the carbon film behind and perhaps ready for the microscope. If necessary, the organic material of the specimen often can be removed using hot concentrated alkalis. This treatment will not affect most preshadowed metal layers. Thus this method has been used successfully in studies of virus particles and macromolecules, providing images of great delicacy and resolution.

Larger specimens which cannot be handled on specimen grids, as well as those requiring strong acids for solution, can be dissolved in small containers after depositing the carbon film. Even concentrated sulfuric acid with permanganate and dichromate added as strong oxidizing agents can be used to clean such carbon replicas. (If permanganate is used, manganese oxides will be formed and must be removed by an additional treatment with concentrated hydrochloric acid). Obviously this treatment also would remove any preshadowed metal films other than those of the noble metals.

Evaporated carbon today is almost the universal choice as the final replica in two-stage replication. If the primary replica consists of a *thin* film of Formvar or Parlodion it is not difficult to dissolve away leaving the intact carbon replica behind. One must perform this operation rather gently, however, and it is wise to begin with solvents that are not too active (start with ethylene dichloride rather than chloroform if the replica is of Formvar, or amyl acetate rather than acetone if of Parlodion). Before beginning, the replica should be cut into squares or pieces less than 1/8 in. wide. They then may be laid across grid screens, plastic side down, and flushed with solvent from below. This can be done on a standard 1×3 in. microscope slide using, alternately, a pipette to add solvent, and filter paper to remove it. If the grids are allowed to dry once or twice during the procedure, the carbon films will settle down flatly upon them and finally adhere to them. Then powerful solvents can be used for the final cleaning.

Difficulties begin when primary replicas are thick. Plastics are prone to swell as they soften, and before going into solution. This is apt to tear apart a delicate carbon replica. Both Formvar and Parlodion are bad in this respect. If a thick primary replica of one of these substances must be used, it usually is necessary to put a supporting film on the carbon until the primary replica is dissolved. Thus, taking advantage of their different solvents, a thin layer of Parlodion can be used to protect the carbon film while a Formvar replica is being removed. In this circumstance, no attempt is made to mount the replica on a grid screen until finally the supporting film is to be dissolved. Waxes can be used similarly.

In this writer's limited experience, resins do not swell badly before going into solution. Therefore, if a thick primary replica is required, there is not serious difficulty in dissolving it without damage to a thin carbon secondary replica. Resins, however, tend to be brittle and will not stand much deformation during mechanical stripping. Therefore they have not enjoyed wide usage.

It will be apparent from the foregoing that although replication is simple in principle, it is widely variable in practice. Certain types of problems can be handled routinely. New problems, however, are apt to pose technical difficulties, and success will depend partly upon a reasonable selection of materials and method, as well as partly upon the ingenuity, dexterity, and perseverance of the investigator.

Literature Cited

Anderson, T. F. (1951). *Trans. N. Y. Acad. Sci.* **13,** 130.

Anderson, T. F. (1953). *Compt. Rend. Prem. Congr. Intern. Microscopie Electronique, Paris, 1950,* p. 567.

Bradley, D. E. (1954). *Brit. J. Appl. Phys.* **5,** 65 and 96.

Mahl, H. (1940). *Z. tech. Physik* **21,** 17.

Schaefer, V. J., and Harker, D. (1942). *J. Appl. Phys.* **13,** 427.

Schmitt, F. O., Hall, C. E., and Jakus, M. A. (1942). *J. Cellular Comp. Physiol.* **20,** 11.

Steere, R. L. (1957). *J. Biophys. Biochem. Cytol.* **3,** 45.

Williams, R. C. (1953). *Exptl. Cell. Research* **4,** 188.

Williams, R. C., and Wyckoff, R. W. G. (1944). *J. Appl. Phys.* **15,** 712.

General References

Bradley, D. E. (1956). Uses of carbon replicas in electron microscopy. *J. Appl. Phys.* **27,** 1399–1412. (The title is apt).

Wyckoff, R. W. G. (1949). "Electron Microscopy." Interscience, New York. (Emphasis upon techniques of replication and shadow casting with much lore that remains valuable today.)

Appendix A

The Literature of Electron Microscopy

Bibliographies

The NYSEM International Bibliography of Electron Microscopy. Published quarterly in "key-sort" card form since 1950 by the New York Society of Electron Microscopists, 2 East 63rd Street, New York, New York. The first 5 years of this bibliography are now available in book form.

Marton, C., Sass, S., Swerdlow, M., VanBronkhorst, A., and Meryman, H., 1950, Bibliography of Electron Microscopy. Natl. Bur. Standards, Circ. 502, U. S. Govt. Print. Off., Washington, D. C.

Published Records of International Meetings

1949

Proceedings of the Conference on Electron Microscopy, Delft. Published in 1950 by Hoogland, Delft.

1950

Comptes Rendus du Premier Congrès International de Microscopie Électronique. Published in 1953 for the Société Francaise de Microscopie Théorique et Appliquée, Editions de la Revue d'Optique Théorique et Instrumentale, 3 et 5 Boul. Pasteur, 165 R. d. Sèvres, Paris, 15e. (Note that this really was the report of the *de facto* second international congress of electron microscopy rather than the first. None was ever designated the "second." The next congress, in 1954, was called the "third.")

1954

Proceedings of the Third International Conference on Electron Microscopy, London, 1954. Published in 1956 by the Royal

Microscopical Society, Tavistock House South, Tavistock Sq., London, W.C. 1.

Rapport Europees Congres Toegepaste Electronenmicroscopie, Gent, 1954. Rijksuniversiteit, Rozier 6, Gent.

Electron Microscopy, Proceedings of the Stockholm Conference (First European Regional Conference). Published in 1957 by Almqvist and Wiksell, Stockholm.

Electron-Microscopy. Proceedings of the First Regional Conference in Asia and Oceania, Tokyo. Published in 1957 by the Electrotechnical Laboratory, Nagata-cho, Chiyoda-ku, Tokyo.

1958

Report on the Fourth International Conference on Electron Microscopy, Berlin. To be published by Springer-Verlag, Heidelburg.

Published Records of Major National Meetings

1956

Proceedings of a Conference on Tissue Fine Structure Held at Arden House, Harriman, New York, Jan., 1956. *J. Biophys. Biochem. Cytol.,* **2,** No. 4, pt. 2 (Supplement), 1956.

1943 to present

The Proceedings of the Annual Meetings of the Electron Microscope Society of America are published annually in the *J. Appl. Phys.*

APPENDIX B

Some Sources of Equipment and Materials*

General sources

Canal Instrument Company, Bethesda 14, Md. (Manufacturers of accessory equipment, mainly electronic, for electron microscopes, particularly adapted to RCA instruments).

Ernest F. Fullam Inc., P. O. Box 444, Schenectady 1, N. Y. (Manufacturer and distributor of specialized supplies and some small equipment for electron microscopy).

Ladd Research Industries Inc., 159 Wagon Road, Roslyn Heights, N. Y. (Another manufacturer and distributor of specialized supplies and some small equipment for electron microscopy).

Specimen grids

C. O. Jelliff Mfg. Corp., Southport, Conn. (Relatively low cost grids as well as screen material of Lektromesh.)

Smethurst High-Light Ltd., Sidcot Heaton, Bolton, Lancs., England. (High quality grids in a variety of patterns.)

Establissements A. Jowa, 105 Rue Bemel, Bruxelles 15, Belgium. (Agents for the Zeefplatenfabriek Veco, Holland, producing high quality grids in a variety of patterns.)

Pyramid Screen Corp., 181 Harvard St., Brookline 46, Boston, Mass. (American agents of the Zeefplatenfabriek Veco, Holland.)

Gunther Frey, Berlin-Britz, Walkenrieder Str. 19.

See also the Ernest F. Fullam Co., and Ladd Research Industries, Inc.

* This list of equipment and materials does not include manufacturers of microscopes and microtomes.

Apertures

Platinum apertures are commonly available from the companies which manufacture microscopes. The Canal Instrument Company also distributes apertures in most common sizes including those to fit their own accessories.

Molybdenum apertures are available from the Aeon Laboratories, Beech Hill, Englefield Green, Egham, Surrey, England. See also the Ernest F. Fullam Co.

Calibration supplies

Ernest F. Fullam Inc., P. O. Box 444, Schenectady 1, N. Y. (Supplies shadowed carbon replicas of calibrated diffraction gratings.)

Physical Research Laboratory, The Dow Chemical Co., 2-280 Bldg., Midland, Mich. (Supplies *gratis* samples of monodisperse polystyrene latexes. Nine different sizes are currently available.)

INDEX

D

E

F

T

U

V

W

X

Z